CW01024529

Adelard of Bath

Louise Cochrane

Adelard of Bath

The First English Scientist

British Museum Press

For Peter

Published by British Museum Press
A division of British Museum Publications Ltd
46 Bloomsbury Street
London WC1B 3QQ

British Library Cataloguing in Publication Data
A catalogue record for this book is available from the British Library

ISBN 0–7141–1748–X

Designed by Andrew Shoolbred
Typeset in Garamond by
Create Publishing Services, Bath, Avon

Printed and bound in Great Britain by
The Bath Press, Avon

Contents

Preface

This book is written for those who, like myself, are interested in the history of ideas: how they are transmitted both in time and space, across the centuries and across geographical boundaries. The writings of Adelard of Bath, and their influence, provide an exceptional picture of how new mathematical ideas helped to develop new approaches to the application of reason and thus to fundamental changes in scientific thought. Adelard's work influenced the development of architecture and even made a contribution to the processes of simple arithmetic.

I first became interested in Adelard at the time when 'New Maths' was being introduced in our schools. Something of the sort must have happened, I thought, when the transition from roman numerals to arabic occurred in Europe. I came across a chance reference to a Benedictine 'monk' from Bath who travelled in Syria at the time of the Crusades, discovered the secret of zero and returned to Europe to disseminate his findings – inaccurate claims, as I was to discover. I was immediately interested, however, in trying to learn more about this. Since I then lived near Bath, I went to Bath City Reference Library and discovered that at the time only one book devoted to Adelard's life and work was available, in German, published in 1935. My next step was to visit the British Library to discover how extensive the available primary sources were.

By now there is an excellent scholarly work in English, published in 1987 by the Warburg Institute, based on a colloquium in which I participated. Individual scholars have devoted in some cases a lifetime to the study of separate aspects of Adelard's work. To present what I hope is a readable summary of all these it has been necessary for me to rely in many instances on secondary sources. For the sake of the general reader I have included explanations which medievalists take for granted.

The very wide range of subject-matter about which Adelard wrote and the extent of his travels in pursuit of more knowledge have given me an impression of the Middle Ages totally at variance with that which I held before. Adelard probably began his career as a clerical member of the Benedictine Order. His father was a tenant of the Bishop of Bath, who was also Abbot. As one of the Bishop's *familia* Adelard had many advantages, and, most importantly, the opportunity of a good education. He was sent to Tours for this and later studied at Laon. He also travelled widely, to Italy, Greece and Syria. Interested in natural phenomena, in mathematics, geometry and the use of the abacus, as well as astronomy and astrology, falconry and music, he wrote on all these subjects.

Adelard first drew on the philosophical ideas, particularly of Plato, which he had absorbed in Tours. Gradually, by extending his interest to Arabic sources, he was able to introduce to the West other Greek ideas as well as some Arab ones, by virtue of his translations from Arabic into Latin. He was more than a translator. His insistence that natural causes could be studied without impinging on theology and that it was essential to assemble and correlate facts as part of one's reasoning process was new to many of his contemporaries.

It is not fanciful to call Adelard 'the first English scientist': he was the key contributor to the conceptual revolution which initiated modern scientific methods. Although his father is thought to have come from Lorraine, Adelard himself was born in Bath, and is thus a true son of Somerset.

I wish to express particular gratitude to those scholars who read this book in typescript and offered comments and advice which greatly enhanced the text: Dr Robert Anderson, Dr Robert Dunning, Dr John H. Harvey, Professor Edward Kealey, Professor Richard Lemay, Professor Angus McIntosh, Dr Tina Stiefel; my further appreciation to the Warburg Institute and those who participated in the Colloquium on Adelard in 1984, especially Dr Charles Burnett, Professor John D. North, Professor Raymond Mercier, Dr Peter Dronke, all of whom have given valuable assistance. At the British Museum John Leopold and Neil Stratford have given additional and useful advice. I would also like to thank Dr David Pelteret and Christopher Hohler for their help with Chapters 1 and 4. Dr J. V. Field introduced me to the intricacies of the astrolabe; the late Professor Eric Forbes encouraged and assisted me with the philosophical background; Rosemary Addison gave useful editorial advice in the early stages. Joanna Champness has been a most considerate and helpful editor at British Museum Press.

Thanks must be expressed to many librarians, particularly at the British Library, but also those in Avon, Somerset and Wiltshire at the start of my research, and thereafter the National Library of Scotland, Edinburgh University Library and Edinburgh Central Library, with its Inter-Library Loan services finding books for me from foreign as well as United Kingdom sources.

M. Jean Jolivet has very kindly given me permission to republish the poem on p. 19 and I acknowledge with appreciation permission to use the quotation on p. 104 from 'The Astrolabe' by J. D. North, © 1974 Scientific American, Inc., all rights reserved; also permission to quote from O. Neugebauer (ed. and trans.), *The Astronomical Tables of al-Khwāriẓmī*, Det Kongelige Danske Videnskabernes Selskab, histor.-filosof. Skrifter, 4,2 (Copenhagen, 1962). The drawings within the text are the work of Christine Barrett.

Any errors which have crept in, however, are entirely my own responsibility. My dedication records my gratitude to my husband for his unfailing patience and sustained interest throughout.

Abbreviations

CEH *Cambridge Economic History from the Decline of the Roman Empire*, ed. E.E. Rich, 7 vols (Cambridge 1952–87)

CHME *Cambridge History of Medieval Europe*, ed. J.R. Tanner, C.W. Previté-Orton and Z.N. Brooke, 8 vols (Cambridge, 1911–36)

DB *Domesday Book*, ed. A. Farley for Record Commission (London, 1783) and *Additamenta*, ed. Sir Henry Ellis (London, 1816), which includes *Liber Exoniensis* (Exon Domesday)

DMA *Dictionary of the Middle Ages*, ed.-in-chief Joseph R. Strayer, 12 vols and index (New York, Chicago, 1982–9)

DNB *Dictionary of National Biography*, compact edn, 2 vols (Oxford, 1975)

DSB *Dictionary of Scientific Biography*, ed. C.C. Gillispie, 16 vols (New York, 1970–80)

EoI *Encyclopedia of Islam*, new edn, ed. H.A.R. Gibb *et al.*, 6 vols (Leiden and London, 1960–90)

EHR *English Historical Review*

FWCR *Friends of Wells Cathedral Report*

OED *Oxford English Dictionary*, compact edn (London, 1971)

PSANHS *Proceedings Somersetshire Archaeological and Natural History Society*

TAPS *Transactions American Philosophical Society* (Philadelphia, Pa.)

WIST *Warburg Institute Surveys and Texts*, ed. Jill Kraye and W.F. Ryan; XIV, Charles Burnett (ed.), *Adelard of Bath, An English Scientist and Arabist of the Early Twelfth Century* (Warburg Institute, University of London, 1987)

List of Illustrations

CHAPTER 1

Early Background

The works of Adelard of Bath, written in the twelfth century, mark a significant stage in the history of ideas.[1] He set out deliberately to acquire for himself, and to pass on, a broadly based understanding of Arabic science, particularly of mathematics and astronomy. He translated the *Elements* of Euclid from Arabic into Latin and thus reintroduced to Europe the full corpus of Euclid's geometry as a logical deductive method. It later became a basic tool of modern science. He also translated the *Zij* (astronomical tables) of al-Khwārizmī and provided the West with additional information about sines and trigonometry which the Arabs had adapted from Hindu astronomy. As well as translating other people's work Adelard was an author in his own right. *Quaestiones naturales*, which attempts an encyclopaedic explanation of natural phenomena, was widely read throughout the Middle Ages. Adelard's emphasis on the study of 'natural causes' and rational enquiry provided another step in the development of modern science. Together with William of Conches and Thierry of Chartres he precipitated a conceptual revolution a century before the new ideas became absorbed into those of Grosseteste, Roger Bacon and Albertus Magnus.[2]

In the course of his career Adelard served both King Henry I and King Stephen. At the outset he is thought to have had an official connection with the Exchequer which was then in its early stages. Later he utilised the astronomical tables and astrological texts he had translated to cast royal horoscopes.

The extensive library holdings of Adelard's writings are ample testimony of his intellectual achievements.[3] Little is known, however, about the man himself. A portrait has to be assembled from the diversity of his interests, from occasional personal comments embedded in his works, and from local archives. He was certainly born in Bath *c.* 1080. He speaks of the city as his birthplace in his treatise on the astrolabe.[4] He considers himself a citizen of Bath and boasts of the fact in *Quaestiones naturales*.[5] It has been suggested that Adelard might have had a wealthy or noble background since in *De eodem et diverso* he casually refers to playing the cithara for the queen.[6] He also shows himself to be extremely knowledgeable about falconry, a sport in which only those of royal or noble blood engaged. He was interested in a great variety of subjects, from falconry, fishing and nature study to mathematics, music and astrology. In later years he gives the impression of flamboyance – wears a green cloak and has a ring, possibly an emerald, set with a *sigil* or seal with astrological significance.[7]

The first occasion on which Adelard, the future scientist, can be identified is when a witness called Athelard, the son of Fastrad (*Athelardus, filius Fastradi*), testifies for Bishop John of Bath in 1106.[8] The spelling of Athelard or Aethelheard with 'ð' or 'th' is usually taken to indicate the Old English form. The name had been a familiar one in Wessex from the eighth century. It is not surprising that Adelard is frequently assumed to have been Anglo-Saxon, especially since he uses occasional vernacular words and refers to Anglo-Saxon practices,[9] but the name could equally have been continental in origin if spelt with 'd' . In this Old German form it was known at Charlemagne's court. To add to the difficulty in establishing Adelard's nationality the spelling with 'th' was Latinised to *Adelardus* before 1120.[10]

The obvious source of further information on Adelard's origin is the Domesday Survey of Somerset in 1087. There is no mention of Adelard but Fastrad is recorded. He was a tenant of Bishop Giso of Wells and thus could have accompanied Giso from Lorraine when King Edward appointed him to the see. The spelling Fast*rad* favours this suggestion. Old English spelling would have been 'Fæstred' or 'Fastred'.[11] According to the Survey Fastrad held 'land of the Bishop, 8 hides in Wells, 5 hides in Yatton, and 1 hide in Banwell'. It was held directly of Giso who remained as bishop during the reigns of both Harold and William the Conqueror. A neighbour of Fastrad's at Yatton was Benzelin, Giso's archdeacon. He continued to serve the next bishop, Bishop John, after Giso's death. Presumably Fastrad could have done so as well.[12] This would have meant that Adelard was born into the bishop's extended *familia*, a fact which could account for the impression he gave of having come from a wealthy or noble family. A bishop's entourage was very like a court. Fastrad appears to have been a tenant rather than an overlord. He probably settled in Somerset during the era of reform and development at Wells. In these times of change his nationality would not have made a significant difference.

Although Fastrad might have come from Lorraine he could have married a local Somerset girl.[13] This would reconcile the two theories about his origins. In 1100 and also in an earlier undated charter at the end of William II's reign another witness called Athelard testified for Bishop John. He was designated as *dapifer*, which means household steward.[14] Adelard of Bath would have been too young to serve in that capacity at this date; it was a senior post with considerable responsibility. Fastrad, however, might well have married a daughter of *Athelardus dapifer*.

Another candidate could have been a daughter of Æthelheard or Othelheard who appears as witness to a diploma (charter) of William the Conqueror in Somerset in 1067. The diploma restores land at Banwell to Bishop Giso and the Cathedral at Wells.[15] The land had been left to Giso for the 'brothers' at Wells by Bishop Duduc to whom it had been given by King Cnut. Various charters reveal that Giso had persuaded Edward the Confessor to confirm grants of Church land which Bishop Duduc had left to Wells, but Harold had reclaimed the land at Banwell on the grounds that Duduc had no right to pass it on. Æthelheard, one witness in a long list, is not given a title or described in any way, but he could have been a local man called upon as a representative of the shire.[16] It is possible that Fastrad married into this family. Further information may result from investigations into local history. It is to be remembered that Fastrad held land at Banwell of the bishop. Although the name, Fastrad, could have been either Old English or Old German, the fact that there is no record of him in the time of King Edward, or as the Domesday Book puts it, TRE (*tempore regis edwardi*), suggests that he probably did come from the Continent.

Although it can quite safely be said that Adelard was entitled by birth to be considered a true son of Somerset, no wealthy family has been identified. His social connections were probably owed to his patron and his later name, Adelard of Bath, a description of patronage as much as of origin.[17] There was at this period in Somerset a falconer called Siward who had served King Edward and continued so doing for William the Conqueror. A serjeanty was established which lasted 300 years. The land involved was more extensive than Fastrad's.[18] The close connection between Bishop John and both William II and Henry I, with hunting rights available to the bishop, would have provided Adelard with opportunities not available in the ordinary way. As part of his alternative background, there may have been additional wealthy connections in Lorraine.[19]

There are several further places, other than in Adelard's own writings, where identification is possible. Adelard is thought to have witnessed again for Bishop John in 1122.[20] He cast a horoscope for Henry I in 1123.[21] He is undoubtedly 'Adelardus de Bada' recorded in the Pipe Roll for 1130, where he is excused a fine by the sheriff of Wiltshire. This particular reference to him by name in official records has been cited as indicating a role at the Exchequer.[22] He is also thought to be 'Adalard de Bath'nian' who witnessed for King Stephen between 1135 and 1139.[23]

A number of candidates bearing the name in some form have been eliminated in this process of establishing that Adelard of Bath was the son of Fastrad, a member of Bishop John's extended *familia*, with connections at court. He was too young, as has already been said, to have been Æthelheard or *Athelardus dapifer* or Adelhard of Liège at Waltham Abbey, mentioned below. The presence of Adelard texts at the Abbey might, however, mean some family association.[24] Nor, as has sometimes been thought, was he the author of a life of St Dunstan in the eleventh century or a Benedictine monk, although he was probably educated under Benedictine auspices. This was a period when, according to Sir Richard Southern, clerical background was of particular importance, since the clergy had a monopoly of all those disciplines which not only determined the theoretical structure of society but provided the instruments of government.[25]

We can now turn to the events which probably formed the background to Adelard's childhood and youth. What seemed an appalling and irreparable disaster occurred in Bath when Adelard was about eight years old. This was the burning of the city in 1088 during a rising led by Robert of Mowbray in an attempt to place Robert of Normandy on the throne instead of William Rufus. There must have been many who feared first for their lives and then for their future. In fact the disaster led to the circumstances which provided Adelard with his opportunity to pursue a learned career. This was because John of Tours, newly appointed Bishop of Wells, was granted by William II the right to purchase the city of Bath from the Crown and to move the seat of the diocese from Wells to Bath. The secular part of Bath had come into the possession of the Crown through Edith, wife of Edward the Confessor. Since the abbot at Bath had recently died, Bishop John was also able to become abbot as well as bishop and the monastery became a cathedral priory. Neither the canons at Wells nor the Benedictine monks of the Abbey of St Peter had an opportunity to give their opinion. John dispersed Giso's semi-monastic community and destroyed its buildings, putting the canons' estates under the control of a provost, his brother Hildebert.

The removal of the diocese to Bath was supposed by historians to have been in response to a decree in council formulated by William I (1075) whereby bishops' seats should be moved as soon as possible from small towns and villages to more populous places. A quite different strategic reason for the transfer of the diocesan seat might have been that William Rufus wanted the city's defences reinforced after the uprising against him on behalf of his brother Robert. Wells was far from being a backwater. Giso had instituted many reforms and new buildings had been constructed. Wells was, however, overshadowed by Glastonbury and did not have the ancient reputation that Bath had. It is thought also that John, who had been a physician at William II's court as well as a chaplain, was interested in the therapeutic possibilities of Bath's mineral springs. He was a learned man and liked the company of educated men. Bath was nearer to Worcester and the great monasteries of the Severn Basin than Wells and thus nearer to the centre of an emerging 'intellectual renaissance'. The opportunity of combining the role of bishop with that of abbot and replacing the monks at the Priory of St Peter with men of his own choice encouraged him further. Indeed, an earlier abbot had become Archbishop of Canterbury. John brought scholarly men to the Benedictine community and later restored to it the property he had taken over. The charter which Adelard witnessed in 1106 was one in which John made arrangements for this. In addition to building a new cathedral John founded a school and established a hospital. He also built a palace for himself.[26]

Coming, as he did, from Bath during this period of growth and expansion meant Adelard had great advantages. His boyhood in Bath was spent while William II still reigned and Bishop John was engaged on his extensive building programme. Little remains now of Bishop John's work. The cathedral church was modelled on that dedicated to St Martin in Tours. Its crossing piers are at the east end of the present abbey. A fragment of arch is encapsulated in the upper parts of the Tudor church. The foundations of Norman nave and aisles lie beneath the floor. Excavations in 1979 revealed some of the north-eastern part of the apse. Bishop John's church extended to the modern Orange Grove roundabout. It was much larger than its successsor and when the seat of the bishopric returned to Wells, it clearly became impossible for the monks to maintain it. It had to be rebuilt in the time of Henry VII. The tower of Bishop John's palace lasted until 1540.[27]

As a physician John of Tours promoted Bath as a centre of healing. Only the King's Bath, thought to have been built for King Henry I, still exists, an eighteenth-century structure on medieval foundations, to give some idea of the scope of the previous building. Adelard would have been able to observe a great deal of the building activity. His writings later demonstrate diagramatically his awareness of how heights are measured for the construction of towers.[28]

Athelardus, filius Fastradi, who testified for John, was described as one of John's knights (*milites*) or officials (*ministri*). His background would have been lay if he were a knight but clerical otherwise. Adelard, the future scientist, completed the trivium and the quadrivium, the basis of all higher education at this time, in Tours; he describes this in *De eodem et diverso*.[29] He could not have gone there without the sponsorship of his bishop.[30] Since John himself came from Tours it is extremely likely that Adelard was sent there by him as a promising young scholar. In *De eodem* Adelard describes the incident when he was called upon to play the cithara for the queen. Queen Matilda is the only likely candidate and the probable date is 1105 when Matilda accompanied her husband to France.[31] Bishop John had been present at Matilda's coronation and it is very likely that Adelard's role as his protégé accounts for the entrée to court circles.

If Adelard followed the customary path to a career in the rapidly expanding administrations of both Church and State, he started out as a clerk.[32] It did not necessarily lead to the taking of holy orders, although this might have been the original intention. Attending a school run by a monastery or a parish did not mean the pupil was intended for the cloistered life or for a parish priesthood; it meant that he there acquired the rudiments of education, how to read, write, count and understand Latin. Those destined for full membership in a religious order were taught within the monastery. To cope with the many different requirements in a diocese cathedrals founded separate schools. Since bishops were increasingly employed for functions unrelated to the Church, these schools became a training ground for court officials. The development of the cathedral schools in Europe, stemming from that founded for Charlemagne by Alcuin of York at Tours and a later one by Fulbert at Chartres, meant that there were a number of schools which attracted intelligent and ambitious young men seeking a higher degree of learning than that available to them locally. Adelard could first have attended the school established by Bishop John and later, when he sought opportunities for higher education, what could be more logical than that he should be sent to Tours where Bishop John had received his own education.

As far as the Church was concerned Europe at this time formed a single cultural unit. It would not be difficult for Adelard to move from a school run by the Benedictines in Bath to a cathedral school at Tours. The expense would devolve upon his family, although some of the living expenses might have been undertaken by the Order. It has proved impossible to discover how Adelard managed to finance all his extensive travel and research over many years. His family and Bishop John must have funded the earlier stages.

John of Tours was a very wealthy man who created opportunities for members of his immediate family. His brother (or brother-in-law) Hildebert became *dapifer*. It is not known whether Adelard continued to have an interest in the Church land of which Fastrad was tenant. Hildebert's son, who became archdeacon, appropriated a good deal of property for his own family's use. On his deathbed he repented and asked his brother Reginald to return it to the Church at Wells. When he did so in 1159 Bishop Robert arranged new prebends to ensure that it was not granted away from the Church again. One of these was Yatton.[33]

The influences at work in the cathedral schools of the period were of paramount importance in training Adelard for his investigations into mathematics and astronomy and other scientific matters. The new interest in mathematics a century before had been inspired by Gerbert of Aurillac who was master of the school at Rheims and ended his career as Pope Sylvester II (999–1003). He had as his pupil Fulbert who founded the school at Chartres. Gerbert helped to introduce the astrolabe to Europe. He spent some of his early years in Spain and this influence can be traced in his use of the abacus with techniques related to hindu/arabic numerals in a form used in Spain.

Gerbert's influence was known to have been of importance in Lorraine (as Lotharingia came to be called) where he had correspondents interested in mathematical ideas. The original point of contact between Arabic science and the Christian West was the result of Carolingian interest in manuscripts to be found in Cordova. Apparently there had been a cluster of Spanish monks at Lyons who had maintained links with a Spanish monastery before the Muslim invasion and there were later occasional relations between Saxon Germany and Muslim Spain. The most significant development before Gerbert's time, however, was when John of Gorze first travelled to Monte Cassino and then Otto I sent him to Cordova as ambassador in 954. There he met a Spanish Jew who understood Latin and was acquainted with Arabic. John probably brought back manuscripts as he had done when he returned from Italy. Gerbert, in fact, could have obtained Arabic manuscripts from Gorze. The Institut du monde Arabe in Paris has in one of its collections a 'Carolingian' astrolabe believed to have been produced in Europe at the end of the tenth century. It was apparently made in Catalonia by a Muslim astronomer for a Christian to have engraved. The symbols used on its circumference for the measurement of degrees reflect a possible transitional phase in the introduction of arabic numerals into Europe.[34]

How did these new ideas reach Bath and the west of England so that a boy with natural mathematical ability could be adequately prepared for advanced scholarship? From the tenth century onwards continental ideas had circulated with English ones, a movement initiated by the reforms of St Dunstan. Later, thanks to King Cnut's apparent preference for churchmen from Lorraine, a succession of scientifically inclined prelates arrived. Lorraine was the European centre from which Arabic scientific knowledge, together with the use of the astrolabe and the abacus, radiated to other countries – to Germany, France and to England. Mathematics was emphasised in its schools. In 1033 Duduc was appointed to the see of Wells by King Cnut. Bishop Giso, his successor, who introduced many reforms and a new constitution, was another Lotharingian, this time appointed by King Edward. Earl Harold's appointments continued the trend, with Walter as Bishop of Hereford (1060–79) and the invitation to Adelhard of Liège to become head of the college of canons he established at Waltham Abbey.

William the Conqueror was similarly anxious to encourage men of known mathematical talent. He invited Robert Losinga (*de Lotharingia*) to his court, a man who was well known for his work with the abacus and who became Bishop of Hereford in 1079. Bishop Odo, William's half-brother, sent Thomas of Bayeux to Liège to increase his knowledge of science before he came to York as Archbishop in 1070.[35]

After the Norman Conquest there were new overlords and the interchange of ideas escalated. Developments in technology produced in men of learning, according to Alistair

Crombie, 'a mentality interested in finding exact experimental answers to practical questions, and so predisposed them to temper their devotion to "abstract generalisation" by observation and measurement'. For instance the need for an accurate calendar to determine the date of Easter was the chief reason for early interest in astronomy. Similarly advances such as the application of water power to industrial processes like cloth-fulling and saw-milling, the development of the windmill (an English invention), and the variety of structural devices used in building new cathedrals illustrate the increasing concern to seek practical solutions by means of experiment and the necessity for accurate measurement and precision. The Domesday Book records the existence in England and Wales of 5,624 water-mills, all but two per cent of which have been located. There were thirty mills along ten miles of the banks of the river Wylie in Wiltshire. In Somerset two mills were paying rent in blooms of iron (lumps or balls of iron squeezed or hammered out in iron-making), so were thus being used as forges rather than to grind corn.[36]

The ideas which most affected Adelard travelled to the west of England, not so far as is known through Giso but through Robert, Bishop of Hereford, and Walcher, Prior of Malvern. Robert had been one of the most distinguished scholars of his day, an abacist, the author of several astronomical works and interested in astrology. After coming to England he served as a royal clerk in William Rufus's court. He became Bishop of Hereford in 1079 and died in 1095.[37] Walcher became interested in astronomical observation after experiencing the darkness of an eclipse in Italy and then discovering on his return to Malvern that the selfsame eclipse had been observed in his own monastery at a different time of day. He commenced his observations in 1090 using roman numerals and fractions but later changed to degrees, minutes and seconds, employing an astrolabe. He learned his new methods from Henry I's physician Petrus Alfonsi, a converted Spanish Jew for whom Alfonso I of Aragon (VII of Castile) had been a godfather at Huesca in Aragon (see Chapter 8). Interest in mathematics and astronomy had therefore been well established in the west of England before Adelard reached maturity.[38]

It should not be assumed that there was a universal enthusiasm about new scientific ideas. Adelard's contemporary, William of Malmesbury, was frightened by mathematics. He describes how in Spain 'Gerbert surpassed Ptolemy with the astrolabe and Alcandraeus in astronomy and Julius Firmicus in judicial astrology. There he learnt what the singing and flight of birds portended, there he acquired the art of calling up spirits from hell .... There is no necessity to speak of his profession in the lawful sciences of arithmetic, astronomy, music and geometry which he imbibed so thoroughly as to show they were beneath his talents and which with great perseverance he revived in Gaul where they had for a long time been wholly obsolete'.

William then goes on to state that Gerbert was probably the first to seize the abacus from the Saracens but that he gave rules which could be scarcely understood by the most laborious operators. He ends by accusing Gerbert of witchcraft and tells a story of how as Pope he had agreed to be under the devil's dominion forever. Gerbert's mathematical ideas were at the root of these accusations. Mathematics was 'dangerous Saracen magic' and Gerbert was a sorcerer.[39]

Adelard would in due course meet opposition to his ideas. Tradition exerted a strong grip. On the other hand he was extremely fortunate in being able to travel widely and seek those teachers who would assist him. The Church certainly did not discourage those who were interested. After the recapture of Toledo Arabic texts were more readily available in Spain. Peter the Venerable of Cluny despatched scholars to undertake the work of translation. Additional impetus to scientific study in Europe came from Salerno which had become a leading scientific and medical centre. The Salernitan school had its beginnings in the eleventh century. In 1065 Constantine Africanus, a Muslim physician converted to Christianity, arrived at the Benedictine Abbey of Monte Cassino and began translations of Galen and Hippocrates. Salernitan physicians began now to produce their own medical literature. It is quite understandable that Adelard, with a background associated with John of Tours, would have been drawn to Salerno, but his interests also took him even further afield. When he wrote *De eodem et diverso*, usually taken as his earliest work, he had left Tours, and travelled to Greece as well. He was a true 'gyrovagus', a wandering scholar. This work provides an excellent summary of early twelfth-century education and helps us to understand the preparatory steps to his later intellectual achievements.[40]

Notes

1 Adelard's significance in this regard is stressed by C.H. Haskins, *Studies in the History of Mediaeval Science*, 2nd edn (London and Cambridge, Mass., 1927), pp. 20–42 and *passim*. The most comprehensive recent work is Charles Burnett (ed.), *Adelard of Bath, An English Scientist and Arabist of the Early Twelfth Century*, *WIST* XIV (London, 1987).

2 Alistair Crombie, 'Some Attitudes to Scientific Progress', in *Science, Optics and Music in Medieval and Early Modern Thought*, (London and Ronceverte, West Virginia, 1990), ch. 2, pp. 23–40 (31); Tina Stiefel, *The Intellectual Revolution in Twelfth-Century Europe* (New York, 1985), pp. 53, 68.

3 Adelard's works will be discussed in detail in subsequent chapters. For complete list see *WIST* XIV, Catalogue, pp. 163–96.

4 Date of birth 1080+, Margaret Gibson, 'Adelard of Bath', *WIST* XIV, p. 7; Adelard of Bath, *De opere astrolapsus*, Cambridge, Fitzwilliam Museum, McClean MS 165, f. 82v. *WIST* XIV, Catalogue 42; see also Louise Cochrane, 'Adelard of Bath and the Astrolabe', *PSANHS* 124 (1980), p. 141. The astrolabe diagrams from the McClean manuscript are reproduced therein on pp. 147–50. One of these is appropriate to the latitude of Bath.

5 Adelard of Bath, *Quaestiones naturales*, 'Die Quaestiones naturales des Adelardus von Bath', ed. M. Müller, *Beiträge zur Geschichte der Philosophie (und Theologie) des Mittelalters* 31, 2 (Münster, 1934), ch. 33, p. 34; ed. H. Gollancz in *Dodi ve Nechdi* (Oxford, 1920), p. 122; *WIST* XIV, Catalogue 29.

6 According to Gibson Adelard's social status is indicated by his knowledge of falconry, a nobleman's sport, detailed in *De cura accipitrum*, *WIST* XIV, Catalogue 4; Gibson, *WIST* XIV, p. 8; also in an incident in *De eodem et diverso*, ed. H. Willner, 'Des Adelard von Bath Traktat', *Beiträge zur Geschichte der Philosophie (und Theologie) des Mittelalters* 4, 1 (Münster 1903), pp. 25, 26, 29; *WIST* XIV, p. 69, passage trans. Charles Burnett, p. 72; *WIST* XIV Catalogue 7.

7 *Quaestiones*, Gollancz, ch. ii, p. 96.

8 W. Hunt, *Two Chartularies of the Priory of St Peter at Bath*, Somerset Record Society 7, 2 vols, (1893), vol. 1, no. 53 (1106).

9 Anglo-Saxon practices are described in *De cura accipitrum*, *De eodem*; Anglo-Saxon words are used in *Mappae clavicula*, *WIST* XIV, Catalogue 25.

10 P.H. Reaney, *A Dictionary of British Surnames* (London, 1958), p. 2; W.G. Searle, *Onomasticon Anglo-Saxonicum* (Cambridge, 1897), p. 39. Searle refers to Athelheard,

Adelard (*c.* 1125) of Bath, scientific writer, implying spellings are interchangeable. The diagraph 'th' or 'ð' is used in early charters. See also P.H. Sawyer, *Anglo-Saxon Charters, An Annotated List*, Royal Society Guides and Handbooks (1968), 'Aethelheard', nos 1177, 1252, 1681; King of Wessex, nos 93, 253–5, 1676; Archbishop of Canterbury, nos 132, 1258, 1259.

11 *DB* (1783), f. 89 r. – v.; *Additamenta* (1816) (*Liber Exoniensis*), ff. 145–7. Henry Sweet, 'The Oldest English Texts', *Early English Text Society* 83 (1885), Glossary, pp. 461–651. On pp. 603–5 is a list of names with *ræd* as the second element beginning with Ælfred (Alfred), but the list does not include Fastred or Fæstred as it should be in Anglo-Saxon. Fastrad would certainly be the corresponding form and would certainly be Latinised as *rad* just as the Old English ræd is Latinised as *red.* Sweet, p. 473, gives personal names with the adjective *ætele* 'noble' as a first element. Among these is Ethelheard along with Æ variants. Those ending *heard* are listed on pp. 485, 486. The *heard* in this is sometimes (in Old English) spelt as *hard*, and *ard* is sometimes found. This can therefore be regarded as just as likely to be English as continental. I am indebted to Professor Angus McIntosh and to Dr David Pelteret for helping me to establish that 'Fastrad' is more likely to have been an Old German name from its spelling and 'Athelard' or 'Æthelheard' could have been either.

12 For Giso and Benzelin see J. Armitage Robertson, 'Some Early Somerset Archdeacons', *Somerset Historical Essays* (British Academy, London, 1921), p. 73.

13 Gibson, *WIST* XIV, p. 8 and n. 6. For further background to the names see Olaf von Feilitzen, *The Pre-Conquest Personal Names of Domesday Book* (Uppsala, 1937),(Æðelheard), p. 184, (Fast ...), p. 250; T. Forssner, *Continental Germanic Personal Names in Old and Middle English* (Uppsala, 1916), p. 8; E. Förstemann, *Altdeutsches Namenbuch* (Bonn, 1900), cols 170, 171 (for Adelard), 501 (for Fastrad).

14 Hunt 1893, vol. 1, nos. 34 (no date), 41 (1100); vol. 2, no. 844, p. 169, another version of no. 34. Gibson, *WIST* XIV, p. 7, also n.4. Gibson does not differentiate between the appellations '*dapifer*' and '*filius Fastradi*' so concludes that Fastrad's son testified four times. Two different men with the name '*Athelardus*' must be involved. For background to legal age see E.J. Tardif, *Coutumiers de Normandie* (Paris, 1881, reprinted Geneva, 1977); for the role of *dapifer* see *Dictionary of Medieval Latin from British Sources*, Fascicule III, D–E, prepared by R.E. Latham and D.R. Howlett (London, 1986); F.M. Stenton, *The First Century of English Feudalism 1066–1166* (Oxford, 1932), pp. 69–78.

15 David A.E. Pelteret, *Catalogue of English Post-Conquest Vernacular Documents* (Woodbridge, Suffolk, and Wolfeboro, New Hampshire, 1990), part 2, 'Royal Charters' 11, pp. 53, 54, Whitsuntide, 1068. Diploma of King William I granting at the request of Giso (gen. Gisonis), bishop (of Wells), thirty hides of land at Banwell (*Banawelle*), (Somerset); Historical Manuscripts Commission, *Calendar of the Manuscripts of the Dean and Chapter of Wells*, 2 vols, (1907, 1914), vol. 1, p. 431.

16 F.M. Stenton, *William the Conqueror and the Rule of the Normans* (London, 1908), pp. 412–14, refers to English witnesses who are present together with Normans. F.A. Dickinson, 'The Banwell Charters', *PSANHS*, N.S. 3 (1877), part 2, p. 49. Dickinson does not give background for Æthelheard but states (p. 53) 'Among the English signatories we have a good many local men whose names appear in the sale of Combe and who are naturally called on to witness a document affecting their own shire ... '. See F.A. Dickinson, 'The Sale of Combe', *PSANHS*, N.S., 2 (1876), part 2, pp. 106, 119. Only some of the names are duplicated, not that of Æthelheard.

17 Letter from Dr Robert Dunning to Louise Cochrane, 7 November 1990.

18 Robin and Virginia Oggins, 'Some Hawkers of Somerset', *PSANHS* 124 (1980), pp. 51–3.

19 Gibson, *WIST* XIV, pp. 8, 13.

20 Hunt 1893, vol. 1, no.54 (1122).

21 North, *WIST* XIV, pp. 158, 159; Catalogue 65, p. 183. See below, ch. 10.

22 *Adelardus de Bada* in Pipe Roll, 31 Henry I (1130), ed. J. Hunter, pp. 22, 32. Reginald Poole, *The Exchequer in the Twelfth Century* (Oxford, 1912), p. 56; Judith A. Green, *The Government of England under Henry I* (Cambridge, 1986), p. 161, suggests 'royal favour' rather than official position.

23 H.A. Cronne and R.H.C. Davis (eds), *Regesta regum anglo-normannorum 1066–1154*, 4 vols

(Oxford, 1968), vol. 3, p. 282, no. 764. See Burnett, *WIST* XIV, p. 4, n. 25. Witness to charter of King Stephen mentioned by E.J. Kealey, *Roger of Salisbury: Viceroy of England* (Berkeley, 1972), p. 49, n. 76.

24 Burnett, 'Introduction', *WIST* XIV, p.4. n. 25.

25 Burnett, 'Introduction', *WIST* XIV, p. 3, n. 20, 21. Sir Richard Southern, *Western Society and the Church in the Middle Ages*, vol. 2, The Pelican History of the Church, ed. O. Chadwick (Harmondsworth, 1970), p. 38.

26 For biographical information on Bishop John see W. Hunt, 'John of Villula', *DNB*, p. 1084; Antonia Gransden, 'The History of Wells Cathedral c. 1090–1547', in L.S. Colchester (ed.), *Wells Cathedral. A History* (Open Books, 1982).

27 W. Rodwell, *The Archaeology of the English Church* (London, 1981), pp. 24, 25; Robert Dunning, *Somerset and Avon* (Edinburgh, 1980), p. 34.

28 See diagrams in *De eodem*, *De opere astrolapsus.*

29 *De eodem*, p. 4; Charles Burnett, 'Adelard, Music and the Quadrivium', *WIST* XIV, pp. 69, 72 and n. 15. The *trivium* and the *quadrivium* made up the seven liberal arts: grammar, rhetoric and dialectic for the *trivium*, arithmetic, geometry, astronomy and music for the *quadrivium.*

30 Helen Waddell, *The Wandering Scholars*, 6th edn (London, 1932), p. 162.

31 Kate Norgate, 'Matilda' (wife of Henry I), *DNB*, p. 1347.

32 Jean LeClerq, *The Love of Learning and the Desire for God* (London, 1974), pp. 238, 239; Sir Richard Southern, 'The Schools of Paris and the School of Chartres', in R.L. Benson and G. Constable (eds), *Renaissance and Renewal in the Twelfth Century* (Oxford, 1982), pp. 113–37; David Knowles, *The Evolution of Medieval Thought* (London, 1962), pp. 80, 84–7.

33 W. Hunt, *The Somerset Diocese, Bath and Wells* (London, 1885), pp. 21–38; Historical Manuscripts Commission, *Calendar of the Manuscripts of the Dean and Chapter of Wells*, 2 vols, 1907, 1914; vol. 1, 1907, 'Ordinance of Robert, Minister of Church of Bath' p. 33. See also Robin and Virginia Oggins, 'Richard of Ilchester's Inheritance: An Extended Family in Twelfth-Century England', paper read on 7 May 1988 at 23rd International Congress on Medieval Studies, Western Michigan University, Kalamazoo, Michigan, with accompanying documentation and genealogies.

34 Marcel Destombes, 'Un astrolabe carolingien et l'origine de nos chiffres arabes', *Archives Internationales d'Histoire des Sciences* 58–59 (1962), pp. 3–45.

35 M.C. Welborne, 'Lotharingia as a Center of Arabic and Scientific Influence in the Eleventh Century', *Isis* 16 (Cambridge, Mass., 1931), pp. 188–99; J.W. Thompson, 'The Introduction of Arabic Science into Lorraine in the Tenth Century (984)', *Isis* 12, (1927), pp. 184–93.

36 Crombie 1990, p. 93; Jean Gimpel, *The Medieval Machine, the Industrial Revolution of the Middle Ages* (London, 1977), pp. 10–14, 175–7. E.J. Kealey, *Harnessing the Wind* (Woodbridge, Suffolk, and Wolfeboro, New Hampshire, 1987), pp. 82–7.

37 Edmund Venables, 'Robert Losinga' (Bishop of Hereford), *DNB*, vol. 1, p. 1245.

38 E.J. Kealey, *Medieval Medicus* (London and Baltimore, 1981), pp. 76–80.

39 William of Malmesbury, *History of the Kings of England*, trans. Revd John Sharpe (London, 1815), p. 199.

40 Michael McVaugh, 'Medicine, History of' *DMA* vol. 8, p. 227; Adelard of Bath, *De eodem*, p. 32.

CHAPTER 2

'Concerning the Same and the Different'

De eodem et diverso is best known for its outline of Adelard's philosophy of non-difference. It is the work of a young man at the beginning of his career when he is attempting to assemble the various ideas he has assimilated in the course of his education. Adelard uses as his excuse for writing the fact that his nephew has made fun of his frequent journeys and accused him of inconstancy of purpose. He wishes to answer his nephew's criticism.

The text is formally dedicated to William, Bishop of Syracuse, a man of considerable mathematical erudition, '*mathematicarum artium eruditissime*'. William may in fact have been the William R thought to have instructed Adelard and others in the use of the abacus.[1] Adelard explains that the greater part of his work will be a dialogue according to philosophical principles between 'Philosophia' and 'Philocosmia', one standing for the Same and the other for the Different. William will find much that is already familiar to him so Adelard makes it clear that his nephew is the intended recipient.[2] On this and other occasions Adelard uses his nephew as a convenient literary device.

Neither William nor Adelard's nephew would require, as we do, an explanation of what Adelard meant by his title. It derives from the theories in Plato's *Timaeus*, a myth about the creation of the universe.[3] In a partial translation by Chalcidius it survived the Dark Ages in Europe and was absorbed into Christian thought. It provides an alternative explanation to that in Genesis of how the world was created by God. Neo-Platonism was an important influence in the cathedral schools, at Tours as at Chartres. Since Plato's views underlie so much of Adelard's work it will be best to summarise them here.

Plato attempts to use what man himself can observe in the heavens and on earth as the basis for the cosmic system he describes. Since God must be an intelligent Being, the method by which the universe was created must be intelligible; Plato assumes that God created the cosmos out of pre-existent and formless matter by devising a gigantic sphere. One circular band around this sphere became the firmament of fixed stars – not fixed in the sense that they do not move, because there is apparent motion of the stars from east to west – but fixed in that the stars remain always in the same relation to each other as the circle in which they are set revolves around the earth. The firmament represents uniformity and comes nearest to perfection in the divine mind. It is called the circle of the

Same. To account for change and diversity there is the circle of the Different, subdivided into seven concentric rings to account for the various cycles of sun, moon and planets. These rings revolve around the earth at the angle of the ecliptic to the earth's central axis and at different speeds. In doing so they provide the reason for all the variety in the universe.

The theory of the rings was improved upon by Eudoxus who altered it into a series of seven concentric spheres within the outer sphere which was the firmament of fixed stars. This view of the universe was accepted by Aristotle. An additional invisible outer sphere was added later by Hipparchus to account for precession of the equinoxes.

In order to accomplish all this the Creator required a pattern or model and this in Plato's view was the real world of Being which existed only in God's mind. Reality was in the Forms or Ideas in the divine mind and outside the universe itself. The world as we would recognise it is Plato's world of Becoming. It is constantly changing and is known to us only by our senses. This world is made up of four elements – earth, air, fire and water – which in the divine mind take geometric form. Earth is a cube, air is an octahedron, fire is a tetrahedron (called a 'pyramid') and water an icosahedron. The cosmos is a dodecahedron. In the universe these elements are associated with one another in different forms of life and matter. The Demi-urge, another name for the Creator, sets all in motion by means of the world soul. It is the world soul which he had made into the circles of the Same and the Different and the world soul which then creates a hierarchy of other souls that generate all life on earth including the lives of individuals.

The circles of the Same and the Different exist also in man, who represents the universe in microcosm. The world soul accounts for two types of revolution in the spherical universe – that of the firmament of fixed stars and contrasted with it the motion of sun, moon and planets at the angle of the ecliptic. Man's soul has two basic forms of judgement which provide him with rational thought: these are affirmation and denial. Something is the same as something else, or it is not. Classification begins. In the divine mind Forms are universals, perfect models for all created things, but in the world itself the many variations from perfect forms are accounted for by the diversity of influences in the constantly changing universe.

Within man as microcosm the powers of reason and decision are located in the head. They represent the divine and immortal part of the soul, but since they exist in a mortal body they are subject to disturbance and consequently error. Mortal parts of the soul are to be found in the heart and the stomach – they account for emotions and appetites.

Since the use of reason and the pursuit of knowledge result from the divine part of man's soul, the study of philosophy reflects man's contemplation of the divine mind. This is why in *De eodem et diverso* Philosophia represents the circle of the Same and Philocosmia that of the Different.

Adelard begins this work by telling his nephew about an experience of revelation which he had had as a student at Tours. He was out late one evening with one of his teachers who pointed out various features of the night sky. The older man then turned to him and said: 'Stay and think things out for yourself. I am going home.' Adelard walked along the banks of the Loire where he was suddenly confronted by two matrons, Philosophia and

Philocosmia, who stood either side of him and began a verbal struggle for the possession of his soul. Each was accompanied by a train of supportive maidens.

In the succeeding dialogue Adelard conveys through these characters his own interpretation of what he has learned during his education. Philosophy's handmaidens are the seven Liberal Arts, the seven divisions of classical scholarship, which in Charlemagne's time had been revived as the basis of education. Their personification as handmaidens of Philosophy is reminiscent of Boethius, who took the idea from Capella, an African grammarian of the fifth century. The beautiful maidens so captured the medieval imagination that they appear often as cathedral sculptures or in medieval art. At Chartres and Laon there are famous sculptures (see Plate 1). A group can be seen at the Victoria and Albert Museum, where there is also a twelfth-century copper casket with champlevé enamel decoration representing the seven Liberal Arts (see Plate 2). The casket was probably used by a wealthy man for his writing implements.[4]

Philocosmia ('Worldliness') represents change, decay and earthly sensualism. She has as her handmaidens Riches, Power, Honour, Fame and Pleasure. Boethius' *Consolation of Philosophy* had dealt with the 'falsely seeming goods' of Fortune which Philocosmia's handmaidens represent. According to Adelard they become oppressed as if with shame and cannot bear the gaze of the seven steadfast Liberal Arts. In the struggle for Adelard's soul Philocosmia on his left hand speaks first.[5] Her persuasiveness is subtle. Why bother with so much frustrated labour in an attempt to define the meaning of the universe and discover a unifying principle in nature? Follow me and my handmaidens can offer you wealth, be it gold, silver or precious gems. They can make you powerful, famous, honoured by your fellow men and provide you with all sensual pleasures. The possibilities offered by Philocosmia are those familiar in the Middle Ages as those offered by Fortune, whose wheel turns. This is a reminder of the inconstancy of riches and power: men may sometimes be on top but may also be ground beneath when they are chained to the wheel.

Philocosmia makes an attempt in the form of a poem to prove false the philosopher's efforts to create unity from the diversity of nature:

> He who has taught what was at first a worthy intellect folly; that it may trust in false appearances of things,
> While all that nature has joined together with her full favour he puts asunder, deceived by the madness of a blind guide,
> Yet also confounds and combines into one species, phenomena, though they be created diverse:
> Let this one be kept at a distance, driven from our shores, to drag his followers with him into darkness.
> Let him, like some Apollo darkling teach in the dark his dark doctrines, and with made-up arguments hold in bondage his like;
> Let him believe none, and be believed by none, as long as his teaching takes from our world nature's true glory.

Philocosmia is challenging the use of reason to determine the true nature of things. Having tempted Adelard with prospects of great wealth and power, she speaks of fame like that of Jason and Hercules and of the pleasures depicted by Epicurus.

Philosophia now has her say.[6] She has seen many difficulties in her rival's presentation and offers arguments for higher pursuits. Philosophia's answer gives Adelard scope for presenting his own ideas. She defends the search for unity in diversity. Adelard explains his reasoning about the development of one cause to several, and from unity to composite, mathematically – in the fact that one can prove the correctness of multiplication by division.

The most important thesis which Adelard presents using Philosophia as his mouthpiece concerns the nature of reality. It was later known as the philosophy of non-difference, an attempt to reconcile Plato and Aristotle in the definition of universals, or at least to suggest they were approaching the same problem in different ways. Aristotle's definition of universals relied on the senses while Plato's did not; Plato's reality was based on Forms in the mind of God rather than in the world of sense experience.

In the early twelfth century scholars had few purely philosophical problems to discuss other than universals. Boethius had offered two solutions, one according to Aristotle and one according to Plato; he had had the original works to study but his successors did not. During the years when Adelard was receiving his education only the *Categories* and *De interpretatione* of Aristotle translated by Boethius were available (known later as the Old Logic).[7]

What is the problem? Briefly it arises from the way in which our minds make classifications. We know the difference between a particular man, John Smith, and man as a human being or man as a species. But how do we know? What is the real distinction which makes us recognise that John Smith is a man but that all men are not John Smith? Aristotle's view, in the simple form in which Adelard conceives it, is that the characteristics which make John Smith a man exist in John Smith, and our minds establish this by comparison with other men we have known, even if the knowledge is intuitive. In Plato's system the universal concept of man exists in God's mind as perfect man or real man; all men whom we know exist because of the pattern in God's mind.

Adelard in *De eodem* explains his conclusion that what we see is at once genus, species and individual, so that Aristotle rightly insisted that universals do not exist except in things of sense. But since these universals, so far as they are called genera and species, cannot be perceived by anyone in their purity without an admixture of imagination, Plato maintained that they existed and could be beheld beyond the things of sense in the divine mind; thus though these men seem opposed they held in reality the same opinion.[8]

This is the view which developed into the philosophy of non-difference. It means that John Smith has within him all the characteristics which are not different from those of other men, as well as those which make him an individual rather than a species.

Historians of medieval thought have underlined the importance of Adelard's comment. Even if he were not the first to apply reason in this way to a metaphysical question, he recorded his interpretation of the problem and attempted to trace the immediate connection between divine ideas and actual being. He agreed with Aristotle that universals were recognised by the senses but continued to believe with Plato that once abstracted by the mind they existed as such in the mind of God.[9]

The philosophy of non-difference forms only a small part of Philosophia's discourse, although it assumed particular importance in the history of ideas. Philosophia also outlines the cosmology of the *Timaeus*, enunciates Plato's theory of the senses' unreliability, and, following Democritus, suggests that matter might be constructed of atoms. She replies poetically herself to the poem of Philocosmia. Her emphasis is that to deny the use of reason is to abandon life to chance.

> Whoever, pretending that the light of his better eye exists not, and who knows not to believe in things insensible,
> Abandoned by the guide reason, where he once excelled, may offer his neck to the yoke of chance.
> Let him possess, but never be master of his possessions, as little bounteous to another as profitable to himself.
> Let him not know the causes or first principles of things nor know himself, ensnared by the love of charming fallacy;
> Let him not know why some stars look balefully on our affairs, while others are benign,
> Nor why the earth stays poised in the centre, lying in equipoise, knowing not to yield to such great natural forces,
> Nor why spring paints the pastures with grass, why autumn fills the homes with corn, and winter congeals the pools with cold.
> Deprived of light, let him seek falsehood instead of truth, arguing that there is no such thing as natural causation.

In the debate between the two matrons, Philosophia triumphs. Her knowledge encompasses a range of sources with which Adelard must have been familiar, such as *Liber Nimrod*, for 'earth in the centre, lying in equipoise'. The intellectual pursuit of philosophy was the supreme work of human intelligence beyond which God's work commenced, hence its association with the circle of the Same and the firmament of fixed stars. Adelard chooses to devote his life to scholarly pursuits and is then enlightened by a vision of the Liberal Arts in procession. Each represents one aspect of philosophy. As subjects they form the trivium and the quadrivium, the two curricula at the cathedral schools. The subjects of the trivium are the first to be represented.[10]

Grammar carries a ferule in her right hand and the alphabet in her left. Grammar proves that man is rational and this distinguishes him from animals. The use of words enables men to find expression for their ideas. The problem of universals appears again – whether reality is in the word or what the word represents.

Rhetoric follows. She has a cheerful aspect and a confident appearance. She has freed men from the unruliness of the wilderness. The foundation of the State is her work. The rightful ordering and well-being of the State are owed to her. Rhetoric represents effective and persuasive public speaking.

Dialectic comes next. In her right hand she holds a serpent and in her left is a tablet of wax with ten divisions. Dialectic is the basis of reason and the means by which men can express themselves without ambiguity. The presentation of Dialectic is related to Boethius' translation of Aristotle's *Categories*. The ten categories of nature are described and the logical method of reducing what is multiple in experience to the unity of systematically organised ideas. Training in dialectic reasoning had become one of the most important aspects of medieval education by this time. The serpent in the personification supposedly warns against being misled by false reasoning. Mâle also states that the careful symmetry in Dialectic's hairstyle represents the balanced terms of the syllogism in medieval sculpture.[11]

Before introducing the second group, representatives of the quadrivium, Adelard makes the point that the trivium relates to the circle of 'Voces', or the names by which ideas are classified.[12] The quadrivium relates to things themselves.[13] This indicates from Adelard's point of view how much more important the quadrivium had become. Arithmetic, whose gown is decorated with numbers, introduces the second group. Here the emphasis is on how all visible aspects of the universe are subject to number. Adelard cites Xenocrates as saying that 'the soul is number moving itself'. He speaks of the importance of number to the Greeks and analyses the differences between odd and even, prime and composite numbers. His account is based on the first part of the first book of Boethius' *De arithmetica.*

Music comes next. She is personified carrying a sweet-sounding tambourine and a little book. Plato's philosophy of the harmony of the spheres is outlined. This was another very popular medieval theme, reflected in art and architecture at Cluny and of great importance in the schools. The thinking on which the medievalists based their artistic representation is summed up by Mâle: 'Ces huit tons où l'on retrouve deux fois le chiffre quatre qui est le chiffre des éléments, les points cardinaux et les sciences du quadrivium, expriment les harmonies de la terre et de l'homme.'[14]

Adelard was particularly fond of music. Putting the words into Philosophia's mouth he incorporates a description of himself (mentioned in Chapter 1) when he played the cithara, a sort of lute, for the queen. He describes how a little boy was so carried away by the rhythm of the music that he waved his arms with great enthusiasm, causing the company to laugh aloud. Adelard suggests that this demonstrated the soul's instinctive response to musical stimulus.[15] As an illustration of a less aesthetic form of response to sound he speaks of an Anglo-Saxon method of ringing bells to bring fish to the surface so they are easier to net.

Geometry is not given special attributes by Adelard. She is usually depicted carrying ruler and compasses. Geometry was certainly Adelard's favourite mathematical subject. He relates her role to the importance of measurement in the maintenance of justice, the prevention of war, the need for land measurement among the Egyptians and its importance to the Greeks. Its importance was proved by the fact that the knowledge of how to take measurements was maintained for generations as a secret of divine authority. In due course geometry had provided men with the ability to measure the height of towers or the depth of wells without tools. Using rudimentary techniques based on right-angled triangles and proportion Adelard explains how these basic measurements can be taken. He gives a brief summary of the practical geometry which had been handed on by the agrimensores (field measurers) for use in land-surveying and followed Roman tradition.

Last to appear in Philosophia's procession is Astronomy. In her right hand she holds a quadrant and in her left an astrolabe. Adelard has reserved these instruments for their more exalted purpose. He has not introduced them for surveying. He explains that Astronomy's science (based on geometry) describes 'the whole form of the world, the course of the planets, the number and size of their orbits, the position of the signs; she traces parallels and colures (these are the great circles which intersect each other at right angles to the poles) and measures with a sure hand the divisions of the zodiac. She is ignorant neither of the magnitude of the stars nor of the position of the poles nor of the extension of the axes. If a man acquire the science of astronomy he will obtain knowledge not only of the present condition of the world but the past and future as well, for the beings of the superior world, endowed with divine souls, are the principle and cause of the inferior world here below'. This passage is similar to one in *Centiloquium Ptolemei* which Adelard himself translated. It reflects his early interest in astrology.[16]

All these summaries have been presented by Philosophy, and Adelard is now, as narrator, exhorted by this formidable matron to pursue in youth the knowledge which will provide solace in his old age. Once he has learned the overall possibilities he must be selective. Philosophy suggests that what is not available in the Latin world can be learned in Greece. In her final comments she remarks on differences of temperament among individuals – the soul exercises its power within each person differently – for some, truth is in the head, for others in the heart and the liver. The same head performs different functions in different areas: imagination in the front, reason in the centre, memory at the back. The left side of the heart has arteries and the right has veins. That which in some men is susceptibility is happiness in others. These remarks reflect the notion that man is a microcosm of the universe and also suggest that Adelard is probably acquainted with Qustā b. Lūqā, *De differentia spiritus et anime*.[17]

Philosophy now departs, leaving Adelard somewhat dazed. He goes on to tell his nephew that he has followed the advice which he had received, travelling widely in order to pursue his studies. He was in fact doing what Petrus Alfonsi recommended in *Letter to the Peripatetics of France*.[18] Adelard continues:

> Up to this point, dearest nephew, I have expounded to you the reason for my journey amongst learned men of various regions ... in order both to clear myself of the unjust accusation and also to convey to you the sense of those studies, so that, when others expound their riches in a complicated way, we may display [our] knowledge. Farewell, and whether I have argued aright, you must decide.

The nephew, we recall, had teased Adelard about his extensive travels and suggested that he lacked concentration. Adelard feels he has justified himself by means of this allegorical description of an experience at Tours. The text also suggests preparation for a role as his nephew's tutor. *De eodem* thus lays the groundwork for *Quaestiones naturales* which is written as dialogue between them. William of Syracuse, the dedicatee, was bishop from 1108 to 1116. This work, therefore, was an early one, although perhaps not the first. If the nephew is old enough to mock him at this stage of Adelard's career, they must have been near contemporaries.

The reference to 'argument' is a reminder of the phenomenal influence of dialectic reasoning at the beginning of the twelfth century. It draws attention to the transition which now takes place with an enhanced role for the quadrivium, influenced in no small part by Adelard himself. There seemed to have been an intellectual explosion in the eleventh century when ideas which had been long dormant suddenly came to life again. Boethius had translated Porphyry, who was responsible for the 'tree' which outlined the procedure on which dialectic analysis was based. Boethius had written: 'Now concerning genus and species, whether they have real existence or are merely and solely creatures of the mind, and if they exist whether they are material or immaterial and whether they are separate from the things we see or are contained in them, in all this I make no pronouncement.'[19]

Porphyry applied to genus and species at least the possibility of Platonic Forms. Boethius speaks of Aristotle's predicaments, substance, accidents and so on, as a matter of words not of things. This is the point which Adelard makes in connection with the trivium – grammar, rhetoric and dialectic. In his view the quadrivium with its mathematical subjects was quite different from the trivium.

The arguments in *De eodem* are not just limited to what Adelard knew of Plato and Aristotle. It has been suggested that his views were a preliminary step in the development of scholasticism and prepared the way for the conceptualism of Abelard, whose name is so much better known.[20] Before either Abelard or Adelard discussed these matters the whole problem of universals had been raised by theologians.

Berengar of Tours had applied Aristotle's analysis of substance and accidents to deny the Real Presence in the Eucharist. Lanfranc of Bec, who had been his pupil, rebutted his argument. In the subsequent quarrel about universals the neo-Platonic realists (among whom St Anselm of Canterbury was the leader) opposed the nominalists who followed Roscelin, who held that the Platonic Form was no more than a word. Roscelin's influence is interesting because he was an independent master and taught neither at a cathedral school nor in a monastery. Yet he had sufficient stature to attract private pupils, among them Abelard.

The problem for theologians was whether the doctrines of Christian faith were challenged by the application of the rules of reason. Roscelin's view that universals were merely words led him to reject the accepted view of the Trinity. The three persons of the Trinity must be so separate as to be three Gods or so united that all three were incarnate in Christ. In St Anselm's view reason was intended to illuminate faith, not to challenge it. Those who fail to discern the reality of incorporeal essences fail before they begin to think. Some faith is necessary before the processes of reason can begin at all. Just as the eye is not altered by the nature of the object presented to it, so the nature of faith is not altered by the scrutiny of reason. As St Augustine had done before him, St Anselm accepted the neo-Platonic Forms and believed they were reflected in genera and species for which they were the eternal models.

St Anselm was the first great thinker of the Scholastic Age. He prepared a carefully reasoned reply to Roscelin on the Incarnation, and followed it later with *Cur Deus Homo* ('Why God Became Man'; see Chapter 3). By the middle of the twelfth century, however, almost all thinkers had adopted one version or another of the Aristotelian abstraction; for none was the Platonic conception of the World of Forms as the only real world and the goal of the mind's endeavour, a reality or even a hypothesis. In so far as subsistent ideas remained they remained as exemplars in the mind of God.[21]

Adelard's contribution to this transitional development deserves to be remembered because it helped to prepare the way for the interest which welcomed the full corpus of Aristotle. Mathematical subjects *were* different; they were not merely concerned with words and classification but with mathematical forms. Heavenly bodies were paramount; hence the importance of the quadrivium. It is also worth pointing out that while discussions based on logic continued in France attention in England was directed now towards the new scientific ideas.[22]

In *De eodem* the views of both nominalists and realists are present. The suggestion that this work is an example of *involucrum*, a literary form used in the Middle Ages to conceal deeper meaning within allegory, means that we should look carefully at some of the ideas with this in mind.[23] In particular Jean Jolivet, who translated the poems of Philosophia and Philocosmia into French, points out that Adelard's Philosophia does not represent divine contemplation but is in fact another form of Philocosmia. She urges the use of reason to study the natural world. The poetry in its French version is much more pleasing than the English translation:

> Quiconque masque le rayon de l'oeil le plus perçant et ne sait croire à ce qui échappe à ses sens,
> La raison qui faisait son honneur cessera de le guider, et il lui faudra tendre le cou au joug pesant de la fortune;
> Posséder sans jouir jamais de ce qu'il possède, généreux pour personne, inutile à soi-même;
> Ignorer et les causes des choses et leurs semences et soi-même tout ensemble, saisi par l'amour d'un mal flatteur;
> Ignorer pourquoi certains astres se cachent à notre vue, pourquoi les autres ne la fuient pas,
> Pourquoi la Terre se tient au centre, immobile, et refuse de céder à l'énorme poids des choses qui la presse,
> Pourquoi le printemps, l'automne, l'hiver, décore, emplit, reserre prés, maisons, ruisseaux, d'herbe, de grain, de glace,
> Privé de lumière qu'il cherche le faux au lieu du vrai, en arguant, que les causes des choses ne sont rien.[24]

We have to consider the possibility that Adelard's thinking in *De eodem* is more advanced than the suggestion that it was his first work justifies. Adelard's interest in natural philosophy is already the dominant motive in his pursuit of learning. There are possible indirect references to Qustā b. Lūqā, as to *Centiloquium Ptolemei* and *Liber Nimrod*. Since *De eodem* provides such a lively picture of Adelard's education at Tours it is logical that our consideration of it has preceded his treatise on the abacus. Following his education at Tours, Adelard returned to Bath for a brief period during which he acted as witness for Bishop John in 1106 and served as a member of the bishop's staff.[25] The next phase of his career took place at Laon.

Notes

1 *De eodem*, p. 1; Reginald Poole, *History of The Exchequer in the Twelfth Century* (Oxford, 1912), p. 52, suggests that Bishop William instructed Adelard in the use of the abacus. Alison Drew, 'The *De eodem et diverso*', *WIST* XIV, pp. 17–24, gives a detailed analysis of the complete text.

2 *De eodem*, p. 4.

3 Plato, *Timaeus and Critias*, ed. Desmond Lee, (Harmondsworth, 1974), pp. 9–17, fig. 5, p. 76; F.M. Cornford (ed.), *Plato's Cosmology, the 'Timaeus' of Plato, translated with a running commentary* (London, New York, 1937), pp. 43, 210, 211; A.E. Taylor (ed.), *A Commentary on Plato's 'Timaeus'* (Oxford, 1928), pp. 360–78.

4 *De eodem*, pp. 4, 5. The literary form of *De eodem* is allegorical and imitates Boethius. See Howard Patch, *The Tradition of Boethius* (New York, 1935), pp. 61, 88, 89; Peter Dronke, *Fabula, Explorations in the Use of Myth in Medieval Platonism* (Leiden, 1974), discusses the use of myth and allegory in the literature of the Middle Ages and the role of *involucrum*. The personification of the Liberal Arts first in literature and then in sculpture is dealt with by Emile Mâle, *L'art réligieux du XIIIe siècle en France* (Paris, 1948), *The Gothic Image, Religious Art in France of the Thirteenth Century*, trans. from 3rd edn Nora Nussey (London, J.M. Dent, 1913, Collins Fontana, 1961), p. 75. Eight figures from the Baptiste Portal of Auxerre Cathedral (plaster casts), 3rd quarter of the thirteenth century, Victoria and Albert Museum. Neil Stratford, 'Metalwork', *English Romanesque Art 1066–1200, Catalogue of the Exhibition* (Arts Council of Great Britain, Weidenfeld and Nicolson, London, 1984), pp. 232–95, no. 287, The Liberal Arts Casket, Victoria and Albert Museum.

5 *De eodem*, Philocosmia, pp. 7, 8, 9. Comment, pp. 38, 39.

6 *De eodem*, Philosophia, pp. 11–17. Comment, pp. 50, 51.

7 John Edwin Sandys, *A Study of Classical Scholarship* (Cambridge, 1903), 2 vols, vol. 1, p. 531, specifies the Aristotelian sources available in the early twelfth century. See also Christopher Brooke, *The Twelfth Century Renaissance* (London, 1969), p. 35.

8 *De eodem*, Philosophy of non-difference, p. 12.

9 Etienne Henri Gilson, *History of Christian Philosophy in the Middle Ages* (New York, 1955), pp. 153, 260; Gordon Leff, *Medieval Thought from Augustine to Ockham* (Harmondsworth, 1958), pp. 87–140, esp. 103, 104, 116, 117; David Knowles, *The Evolution of Medieval Thought* (New York, 1962), pp. 93–115; Guy Beaujouan, 'Medieval Science in the Christian West', in René Taton (ed.), *Ancient and Medieval Science, A General History of the Sciences*, 4 vols, (London, 1963), vol. 1, pp. 468–531.

10 For possible acquaintance with *Liber Nimrod* see ch. 4, n. 9. *De eodem*, Adelard's choice, p. 17. Comment, p. 73. Procession of Liberal Arts, Trivium: Grammar, pp. 18, 19. Comment, pp. 90–3. Rhetoric, pp. 18–21. Comment, pp. 93–5. Dialectic, p. 22. Comment, pp. 96, 97. See Drew, *WIST* XIV, pp. 19, 20.

11 Mâle 1948, p. 78.

12 *De eodem*, Trivium as Voces, p. 13. Comment, p. 98.

13 *De eodem*, Procession of Liberal Arts continued, Quadrivium: Arithmetic, pp. 23, 24. Comment, pp. 98, 99. Music, pp. 25–7. Comment, pp. 99–101. Geometry, pp 28–31. Comment, pp. 101–3. Astronomy, pp. 31, 32. Comment, pp. 103, 104. See Drew, *WIST* XIV, p. 21.

14 Emile Mâle, *L'art réligieux du XIIe siècle en France* (Paris, 1947), p. 321. Charles Burnett, 'Adelard, Music and the Quadrivium', *WIST* XIV, pp. 69–86, gives a detailed analysis of Adelard's musical appreciation and translates the relevant section of *De eodem*.

15 See above, ch. 1, pp. 1, 5.

16 Passage trans. Theodore Wedel, *The Mediaeval Attitude towards Astrology particularly in England*, Yale Studies in English 60 (1920), pp. 49, 50. Richard Lemay has drawn my attention to the similarity of this passage with ps-Ptolemy, *Centiloquium Ptolemei*, trans. from Arabic by Adelard, *WIST* XIV, Catalogue 1, p. 166. See below ch. 9.

17 *De eodem*, man as microcosm of world soul, location of functions in human body, p. 32. Comment, p. 81; Richard Lemay has also drawn my attention to Qustā b. Lūqā, *De differentia spiritus et anime*, *WIST* XIV, Catalogue 5, p. 167.

18 Petrus Alfonsi, *Letter to the Peripatetics of France*, listed in *WIST* XIV, Catalogue 19,

pp. 172, 173; *De eodem*, Farewell to nephew, project for future, p. 34.

19 Porphyry's *Isagoge* is an introduction to Aristotle and dialectic. For background see L.W. Jones, 'Cassiodorus', in Austin P.E. Evans (ed.), *Records of Civilization, Sources and Studies* (New York, 1946), pp. 145–209 (161, 162). See also Knowles 1962, p. 108, *et seq.*

20 J.A. Hauréau, *Histoire de la Philosophie scolastique* (Paris, 1850), p. 345; George Sarton, *Introduction to the History of Science*, vol. 2, part 1 (Baltimore, 1931), p. 168.

21 Sir Richard Southern, *Saint Anselm and his Biographer* (Cambridge, 1963), pp. 55–93. Margaret Gibson, *Lanfranc of Bec* (Oxford, 1978), p.96.

22 Gilson, p. 260.

23 Dronke, *Fabula*, p. 61 ff.; Burnett, *WIST* XIV, p. 75.

24 Jean Jolivet, 'Adélard de Bath et l'amour des choses', *Metaphysique histoire de la philosophie*, Receuil d'études offert à Fernand Brunner (Neuchâtel, 1981), pp. 77–84. Jean Jolivet has kindly authorised the republication of his translation of Philosophia's poetic address. He refers to an '*a-religieux*' aspect of Adelard's thought, pp. 81, 82.

25 See ch. 1.

CHAPTER 3

Tutor at Laon

Early in his career Adelard accompanied his nephew and other pupils to Laon.[1] Universities were not yet formally established in Europe at this time, but popular masters would draw students from many places and cathedral schools developed reputations based on the work of particular men. Laon was an outstanding example of this and became a centre of learning with close English connections. An Italian student writing home said that Laon was very crowded with 'clerks from London'.[2] Roger of Salisbury, Henry I's justiciar, sent his nephews, Nigel (later Bishop of Ely) and Alexander (later Bishop of Lincoln). Ranulf Flambard, Bishop of Durham, sent his clerk, William of Corbeil (later Archbishop of Canterbury). There were many others. It was customary for young men interested in study at a more advanced level to take with them pupils of their own. This is what Adelard is thought to have done.[3]

Two men provided the leadership at Laon. The head of the cathedral school was Master Anselm, not to be confused with St Anselm, who was by now Archbishop of Canterbury. Master Anselm was assisted by his brother Ralph, a mathematician, who later succeeded him. Master Anselm's reputation arose more from the fact that Laon was one of the first secular schools to teach theology than from his personality. The theological teaching was conservative but the method was very advanced, as has been established by hundreds of manuscript fragments which survive. The teaching method increased Laon's reputation during the period when arguments about nominalism and realism were to the fore. Question and answer were used as a means to establish the truth of interpretations of the Bible.

The attempt at Laon to apply dialectic analysis to biblical texts in order to solve theological problems would obviously raise awkward questions. One such is reflected in St Anselm's book *Cur Deus Homo* ('Why God Became Man') written about the time he became Archbishop. St Anselm set up his own logical structure and argued within it as he had done previously with the ontological proof of God's existence. For him reason was an activity of faith. He proved to his own satisfaction that the Incarnation was a fact and a necessary fact. There is one chapter in his book in which he records an argument about the devil identical with one in a Laon manuscript fragment, but with a quite different conclusion. It was at first thought that Ralph of Laon, in whose hand the fragment is written, had produced with his students a reply to St Anselm. Now it is thought that St Anselm was troubled by opinions expressed at Laon which he felt were untenable. When the matter was drawn to his attention he provided his own answer.

The question involved an argument that man's sinfulness had forced God to concede certain rights to the devil and allow him dominion in a limited sphere. The Laon fragment accepted the validity of this argument; St Anselm refuted it. For him the Incarnation rendered the devil powerless. God had made no such concession. By becoming man in Christ, and suffering death upon the cross for man's redemption, God had provided the means by which man could always be forgiven whatever the sin.[4]

Scholasticism at this time was in its earliest phase, both as a way of thought, represented in St Anselm's views, and as a method of presentation which was only beginning to evolve. Dialectic based on those works of Aristotle called the Old Logic (referred to in Chapter 2) would be gradually strengthened by other Aristotelian works as they became known. This was to be the New Logic. Another element was the disputation (*disputatio*). The teaching method at Laon reflected this. There was also the method of *Sic et Non* contradictory statements to be reconciled, a method later associated with Abelard. All this was very much part of the background at Laon to which students were exposed. St Anselm's interest in Laon was very natural because of the number of English clerks who were training there.

Abelard, like his near namesake, Adelard, also spent time at Laon. The two men could have met but it is of course impossible to say whether they ever discussed matters of mutual interest since they never refer to each other. By 1108 they had both formulated their own views on the question of universals. Adelard recorded his in *De eodem*. Abelard had studied with Roscelin. Reacting against the realism of St Anselm's followers Abelard's early formula had been somewhat similar to Roscelin's, that a universal is merely a vocal sound. He changed this later to the universal as a mental word, *sermo*. His interpretation of the two notions is similar to that expressed by Adelard. The mental representation is more indistinct than the thing itself because it omits all that the individual does not share with other individuals and it is within the mind, not outside it. Before coming to Laon Abelard had challenged the extreme realism of William of Champeaux whose view was that the only reality was in the divine mind. William changed this to a later view that the individual contained both genus and species; Abelard said this would mean that both were the same which was ridiculous. Abelard's skill as a logician enabled him to use dialectic argument with devastating power. Neither William nor Adelard would have found it easy to discuss these matters with him. At Laon Abelard became bored with Master Anselm's teaching and having briefly conducted his own school he returned to Paris in 1112 to teach theology himself. His success helped to ensure that the Sorbonne became the first great French university.[5]

For his part Adelard seems not to have become involved in the arguments engendered by Roscelin's nominalism vs. St Anselm's realism where dialectic reasoning established truth for some and obscured truth for others. He does not take further the discussion of universals which he raised in *De eodem*. Along with a number of his contemporaries his interest in the subjects of the quadrivium begins to exceed those in the trivium. To this end he may well have been attracted to Laon by the reputation of Master Anselm's brother, Ralph.

The four mathematical arts were considered to be quite different in form from the arts of language. Ralph was a leading mathematician and the author of an important treatise on the abacus.[6] Adelard wrote a similar treatise called *Regule abaci* while he was in France. In the letter of dedication he reminds his dear friend, 'H', that when they dined together at the philosophical table he tried to slip between the lips of his friend a few morsels from a platter which had four compartments. His play on words (*mensa* means 'course' as well as 'table') implies that the *mensa Pythagorea* (his term for the abacus) would make studies in the quadrivium more palatable. 'H' is perhaps one of Adelard's students at Laon.[7]

A further indication of the significance of mathematical teaching at Laon is that the abacus methods which formed the basis of Henry I's Exchequer were studied there. This was undoubtedly due to the influence of Roger of Salisbury who was in charge of Henry I's administration during the king's frequent absences from England. Adelard was almost certainly known to him. Roger of Salisbury's nephew, Nigel, Bishop of Ely, having received his training in Laon, later sent there his own son, Richard, the author of the *Dialogus de Scaccario*, which became the classic manual of Exchequer practice. Richard ultimately managed the Exchequer for Henry II.

There was increasing familiarity with the abacus in Europe at this time. Another Englishman called Turchil wrote an abacus treatise earlier than Adelard's. He and Simon of Rutland (also called Simon de Rotol) were supposedly both pupils of William R. (thought by Haskins to be the later Bishop of Syracuse). Robert of Lorraine, Bishop of Hereford (1079–95) and previously a king's clerk, was an abacist, as mentioned above. By the end of the eleventh century English scholars were undertaking abacus studies and links existed between such scholars and the royal court.[8]

The author of a medieval teaching book on mathematics which forms the basis of twenty-three manuscripts still in existence came from eleventh-century Lorraine. The book is known as *Boethius' Geometry II*, but modern scholarship has established that the section on the use of the abacus derives from Gerbert and the geometry of the agrimensores as well as from Boethius. In the Roman tradition geometry consisted of the body of measuring techniques known as *ars gromatica* (from *groma*, a surveyor's pole). The author was a poor mathematician and gets a number of calculations wrong. The book demonstrates the low standard of mathematics in Europe in the middle of the eleventh century and enables us to compare the rapid progress in the twelfth. The abacus section is of particular interest because hindu/arabic numerals are illustrated.[9]

It is generally accepted that our present-day numbers are of Indian origin and had been developed by Western Arabs in Spain using Hindu arithmetic and a dustboard abacus. The numerals were called *ghubār* or dust numbers because they could be scratched in dust or sand. The dustboard was one of three types of abacus familiar to the Romans which had also been used by the Greeks. It was particularly useful in geometry. Calculations were made by writing in dust with a stylus called a *radius* in Latin. Numerical calculations were based on a decimal system. The digits 1–9 were represented by the first letters of their

Greek names or by symbols. Their values depended on the position in which they were placed in columns representing units, tens, hundreds, thousands and so on.

The mathematician called Gerbert, who became Pope Sylvester II in the year 999, was born in Aurillac and sent as an unusually clever boy by French monks to a Spanish monastery. Here he was introduced to *ghubār* numbers. Gerbert then developed abacus techniques which were adopted by other mathematicians more by oral than by written teaching.[10] (See Figure 1.)

The use of zero in combination with the nine *ghubār* numbers in a decimal system was not understood in the West until the translation of al-Khwārizmī's instructions on how to calculate with Indian numbers (see below, Chapter 8, pp. 80, 81). In the years following Gerbert a blank disc called a *rotulus* was used to fill a blank column to avoid mistakes and to ensure that the empty column retained its function. Adelard instructs his readers about 'taking down' or 'differences' in accordance with how the null is reached, that is how to use the empty column, later zero. Adelard uses the term *siposcelentis*. *Rota* and *sepos* are found in similar treatises.

The suggestion has been made, plausibly enough, that the arabic numbers employed or listed by many authors on the abacus had originally included a zero, but its usefulness was limited when employed upon a board with fixed columns. The technical terms used in the early days disappeared to be replaced with *cifra*, *circulus* or *theca*. For his abacus Gerbert had 1,000 apices made with symbols for 1 to 9 to use in the columns of a counting board. His columns were grouped in triads to make it easier to place the counters. They were called apices because of their shape. The apex was essential to show the symbol the right way up.

Adelard's treatise and those similar to it were intended to assist students to remember the techniques. The counting board had columns ruled vertically with horizontal lines as well, or a table was covered with a checkered cloth. It looked like a chessboard, the Latin word for which is *scaccarium* from which the word Exchequer is derived. The real value of any counter depended both on the column, and the position within the column in which it was placed. Since addition and subtraction were too obvious to require explanation, the treatises concentrated on methods for multiplication and division.

Adelard's understanding of the abacus is considered a reason for believing that he had an official role in Henry I's Exchequer. In fact little more would be required for this than knowledge of how to add and subtract. The Exchequer counting table was large enough for calculations to be supervised publicly and every manœuvre was closely watched. The table was 3m (10 ft) long and 1.5 m (5 ft) wide. It was covered with a cloth on which the columns were embroidered. The calculations were done according to the money itself, not

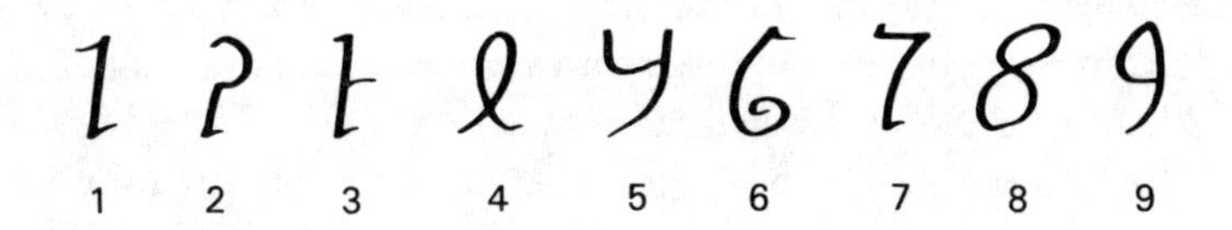

1 Ghubār numerals.

just in tens but proceeding from pence to shillings, to pounds, to scores of pounds, to hundreds, but rarely to thousands. The zero which ultimately made Arabic or Indian numeration more intelligible was indicated by an empty column. It did not require a high level of education to understand the mathematical operations since they had to be as clear to the person watching as to the clerk. The result of the calculation on the Exchequer table was recorded in roman numerals with a dot diagram to show what the position of the counters was on the abacus.

In Adelard's *Regule abaci* the explanations are in accordance with a decimal system but symbols are given for duodecimal fractions since Roman weights and measures were based on twelve ounces to a pound. In this respect the treatise follows Gerbert.[11] Multiplication was fairly straightforward. Division was more complicated. In describing division Adelard and Ralph of Laon refer to a 'golden' and an 'iron' method and also to a method which was a mixture of both. The 'golden' method was *sine differentias* ('without differences'), which seems to have meant that it was possible to compute directly with the figures involved. In using the 'iron' method one rounded up the divisor to the nearest ten or hundred and performed additional calculations with the differences from ten.[12]

The aim of the operator was to reduce any problem to a series of simple calculations and use the abacus to remind him of the answer to each of these until he moved to the next. Only the total end result was recorded.

Gerbert's apices were ultimately abandoned for blank counters which were used so efficiently with counting boards that it took several centuries for roman numerals to become outmoded. The method of calculation was to place a blank disc on the line for five; discs in the space above or below were ones. It was absolutely necessary to have an empty column to indicate zero. A solitary unmarked disc meant one.

The requirements of the Exchequer were more complicated than those of an ordinary counting board. This was because there were columns to represent pennies, shillings and scores of pounds. In the columns requiring numbers up to twenty, a counter above the line to the right meant five and to the left meant ten. In the pence column calculators also devised special arrangements (see Figure 2).

The names for the first nine numbers were Igin (1), Andras (2), Ornis (3), Artoes (4), Quinis (5), Caletes (6), Zenis (7), Temerias (8) and Calentes (9). Early manuscript illustrations give these with their symbols and add also a symbol for *sepos* or *cifra.*[13]

The authors of abacus treatises were mathematicians with an interest in theory. Adelard borrows from dialectic to look for genus and species in number. Digits (*digiti*) were numbers smaller than 10. Numbers like 10 and 20 were *articuli*. He thought of these as points of junction. *Articuli* and *digiti* were combined in composite numbers, *compositi*. When the divisor is a digit the division is called *simplex*. When the divisor is a composite number whose figures fall into adjacent columns the division is known as *composita*. Where the divisor is a composite number whose figures fall into columns leaving one or more vacant (304, 3040) the division is called *interrupta*.

The use of these terms has to be understood also in relation to the way in which numbers were handled traditionally in business transactions. *Digiti* could derive from finger counting and *articuli*, numbers like 10 and 20, were probably shown by manipulating the hand or finger joints in some way. The *articuli* were combined with *digiti* to form higher numbers. To record them formally in a business deal Anglo-Saxons used a tally stick, distinguishing different money values by notching different sections. The stick was then split in two lengthwise down the middle so that creditor and debtor each had a half suitably notched. The tally stick was used to record payments to the Exchequer.[14]

The Latin term *articuli* was used rather freely and sometimes meant all numbers higher than 9 rather than just 10, 20, 30 and so on. The term *differentia* is also a difficult one. It later came to be synonomous with decimal positions on a counting board divided into columns. This did not mean decimal fractions; they came much later. The fractions with

£ 10,000	£ 1,000	£ 100	£ 20	£ 1	s	d
		3×100	3×20	10+4	5+5	3+3
		3×100	2×20	10+5+4	10+3	2
				10+4	10+7	4

Clerk Calculator Cutter of Tallies

2 Exchequer Table. The amounts shown are £374 10s 6d (top), £359 13s 2d (middle), £14 17s 4d (bottom).

Officials of the Court of the Exchequer sat around three sides of the table overseeing the operations performed by the calculator sitting at the centre of the lower side, with the recording clerk on his left and the cutter of tallies on his right. The amount due is shown at the top. After the debtor has presented tallies to indicate what he has already paid, this amount is deducted and the remaining debt shown on the bottom line.

which the abacists familiarised themselves were all fractions with the numerator 1, with the exception of twelfths. Abacists had to know how to set out and use a fraction table. The difficulties they encountered were mainly familiarity with the symbols and technical mastery.[15]

The rules for the Exchequer made it quite clear that its clerks would follow their own procedures and divide their columns in accordance with monetary values, using insignificant coins such as pennies as counters with values expressed entirely by their position in the columns. Later valueless tokens called jettons were manufactured to be used with counting boards. The fact that important financial transactions were conducted in this way helps to explain why arabic numerals did not come into general use for such a long time, even though they were easily understood and familiar at least to mathematicians. Interestingly enough their use was delayed in the Arab world after their introduction from India. Abjad numbers based on the letters of the alphabet continued among the eastern Arabs as did roman numerals in the West. Those accustomed to using a Roman abacus for routine calculations found it quicker to do so than to perform sums in their head.

Regule abaci is a practical book, but Adelard was interested in more than numerical notation. He had shown that he felt that mathematics provided the key to explaining relationships in the universe and would clearly wish to follow up his interests in geometry, astronomy and natural science. He would not wish to spend the rest of his life teaching basic arithmetic. Logical argument using the arts of the trivium rather than those of the quadrivium did not appeal to Adelard as it did to Abelard. Nor does Adelard appear to have participated in discussions of mathematics as theology, as others did. Thierry of Chartres suggested that the laws of mathematics were the same for God as for his universe, that scientific laws were also theological laws. The first law of mathematics considered indispensable to the theologian is the principle that unity gives rise to every kind of otherness. The perfect unity of creation gives rise to all multiplicity.[16] This Adelard took for granted as is shown by *De eodem*. He appears to have left Laon some time before Abelard did and must have felt that he had learned all he could.

The similarity in these two names has led to confusion but intellectually the careers of these two men completely diverged. Two works attributed to Abelard are felt perhaps to have been the work of Adelard because of their mathematical nature – the *Liber sacratus* and *Rhythmomachia*.[17]

Abelard's rather cruel comments about Master Anselm damaged this teacher's reputation. He was undergoing considerable administrative difficulties during the time that Abelard was in Laon owing to the strong reaction of the townspeople against their new bishop. In 1107 at Henry I's behest his former chancellor, Waldric (or Gautry), was appointed to the see and thus became Anselm's superior. Master Anselm did all in his power to challenge this appointment; he even made an appeal to the Pope but all to no avail. Waldric had virtually no qualifications for the post, and, as described by Guibert de Nogent in his autobiography, written at the time, was a singularly greedy and unpleasant character. The account reveals some of the less attractive side of Church appointments and political intrigue. The crisis reached its climax when the bishop had the charter of the commune withdrawn. In the ensuing riots a plot to murder the bishop was organised.

Master Anselm knew about it and warned Waldric but the warning was ignored. After the murder the cathedral was destroyed by fire. Subsequently the canons of Laon travelled to England to try to raise money to rebuild their cathedral.[18]

These events also point to the economic restlessness and change in Europe during the early stages of the dissolution of feudalism. In France the *faubourgs*, former Roman walled cities, had emerged as safe communities where trade could be conducted. Markets and fairs were held regularly on different days and merchants could participate freely, protected by established authority. Originally other *bourgs* were ill-adapted to the needs of merchants, but merchants demanded the same rights as those in the *faubourgs*. When they succeeded, they attempted to establish communes, seized municipal power and sought charters from the king or overlord.[19] The people of Laon had gained such a charter and were determined not to lose it.

Throughout Europe the expansion of municipal law followed the expansion of commerce. In north Italy at the start of the twelfth century fully fledged communes were emerging from the semi-autonomy of the eleventh. From Constantinople and the Levant Italians brought spices, sugar, cotton, rare fabrics, dye woods and wine. Salt was the staple export of Venice. To the Transalpines the Italians handed on oriental and African products and their own cloth, finer qualities of dyed cloth. The trade was vigorously organised, merchant fleets accompanied by galleys sailed twice a year from Venice, Pisa and Genoa. At Constantinople and the Syrian ports there were colonies of Venetians, Pisans and Genoese governed by consuls with storehouses for wines and ships' tackle. The guilds of money changers prospered. Merchants who sold textiles, spices and similar products coupled commercial skill with a thorough knowledge of the goods they handled. One of the essential characteristics of these traders was their mobility. They collected in parties and travelled together by boat or road to transport wheat, wine, wool or cloth to distant places.[20]

Adelard's desire for change may well have been stimulated by the possibility of travelling with a group of merchants. His wish to increase his knowledge would be incentive enough. His ability to handle accounts would be useful in such a group. Trade in books was possible, although a Syrian merchant, Abu-al-Fadl, writing in *The Beauties of Commerce* as early as the tenth century, advised against commerce in scholarly books since the learned men who bought them were mostly poor and few in number.[21] A twelfth-century ordinance in Seville represents another point of view: it forbade selling learned books to Christians in Spain because they 'translate them and ascribe them to their bishops'.[22]

How Adelard travelled and how he supported himself are not known; but before the horrific events of 1115 in Laon, including the murder of the bishop, Adelard had departed to pursue his studies, as he records, 'among the Arabs'.[23]

Notes

1 Adelard states that he parted from his nephew and other pupils at Laon, *Quaestiones naturales*, Müller, p.4.

2 C.H. Haskins, *Studies in Mediaeval Culture* (Oxford, 1929), p. 12.

3 E.J. Kealey, *Roger of Salisbury: Viceroy of England* (Berkeley, 1972), pp. 23, 24; Sir Richard Southern, 'The Schools of Paris and the School of Chartres', in R.L. Benson and G. Constable (eds), *Renaissance and Renewal in the Twelfth Century* (Oxford, 1982), p. 117.

4 Sir Richard Southern, *Saint Anselm and his Biographer* (Cambridge, 1963), pp. 79–87, 357–61.

5 James Bowen, *A History of Western Education*, vol. 2, *Civilization of Europe Sixth to Sixteenth Century* (London, 1975), pp. 50–86. See also George Makdisi, 'The Scholastic Method in Medieval Education – an Inquiry into its Origins in Law and Theology', *Speculum* 49 (1974), pp. 640–61.

6 Ralph (Radulphus) of Laon, ed. A. Nagl, 'Der Aritmetische Tractat des Radulph von Laon', *Zeitschrift für Mathematik und Physik* 34 (1889, Supplement, Leipzig, 1890), pp. 87–196.

7 Adelard of Bath, *Regule abaci*, ed. B. Boncompagni, *Bulletino de Bibliografia e di Storia della Scienze Mathematiche* 14 (1881), pp. 1–134, Latin text p. 91, symbols for fractions p. 109, abacus as *mensa Pythagorea* p. 91; Gillian Evans, 'A Note on the *Regule abaci*', *WIST* XIV, pp. 33–5, provides a summary of Adelard's text and suggests it was written in France.

8 Reginald Poole, *History of the Exchequer in the Twelfth Century* (Oxford, 1912), p. 24; Richard Fitznigel, *Dialogus de Scaccario: The Course of the Exchequer* ed. and trans. the late Charles Johnson with corrections by F.E.L. Carter and D.E. Greenway (Oxford, 1983), pp. xxxv-xxxvii; Kealey 1972, pp.42–50; Judith A. Green, *The Government of England under Henry I* (Cambridge, 1986), pp. 40, 41, 161; Haskins 1927, ch. 15, 'The Abacus and the Exchequer', pp.327–35 (329).

9 Menso Folkerts (ed.), *'Boethius' Geometry II: Ein mathematisches Lehrbuch des Mittelalters*, (Wiesbaden, 1970), pp. 83–9; see also comments in Menso Folkerts, 'Adelard's Versions of Euclid's *Elements*', *WIST* XIV, pp. 60, 61.

10 Guy Beaujouan, 'Medieval Science in the Christian West', in René Taton (ed.), *Ancient and Medieval Science, A General History of the Sciences*, 4 vols, (London, 1963), vol. 1, pp. 468–531, (473).

11 Gerberti, postea Sylvester II Papae, *Opera Mathematica*, ed. Nicholas Bubnov (Berlin, 1899); Harriet Lattin, *The Letters of Gerbert* (New York, 1961), pp. 45, 46; 'The Origin of our Present System of Notation according to Theories of Nicholas Bubnov', *Isis* 19 (1933), pp. 181–94.

12 J.M. Pullen, *History of the Abacus* (London, 1968); M. Mahoney, ch. 5 'Mathematics', in David Lindberg (ed.), *Science in the Middle Ages* (Chicago, 1978), pp. 147, 148, 170.

For those who are interested the method of division by differences works as follows:

To divide 400 by 7

First you round up 7 to 10, difference 3

400 ÷ 10 = 40
40 × 3 = 120
120 ÷ 10 = 12
12 × 3 = 36
36 ÷ 10 = 3 remainder 6
3 × 3 = 9 + 6 = 15
15 ÷ 10 = 1 remainder 5
1 × 3 = 3 + 5 = 8
8 (7 + 1) ÷ 7 = 1, remainder 1

Add together 40 + 12 +3 +1 +1, remainder 1 = 57, remainder 1

Our method: 7 into 400
7 into 40 = 5, remainder 5
7 into 50 = 7 remainder 1
answer = 57, remainder 1

13 Richard Lemay, 'Arabic Numerals', *DMA*, vol. 1 (1982), pp. 382–98; 'Roman Numerals', vol. 10 (1988), pp. 470–4; 'The Hispanic Origin of Our Present Number Forms', *Viator* 8 (1977), pp. 435–62; M. Mahoney, 'Mathematics' in *DMA*, vol. 8 (1987), pp. 205–22.

14 Gillian Evans, 'Schools and Scholars, The Study of the Abacus', *EHR* 94 (1979), pp. 71–89; 'Theory and Practice in Treatises on the Abacus', *Journal of Medieval History* 3 (1977), pp. 21–38; Solomon Gandz, 'The Origin of the Ghubār Numerals or the Arabian Abacus and the Articuli', *Isis* 16 (1931), pp. 393–424.

15 Frances Yeldham, 'Fraction Tables of Hermann Contractus', *Speculum* 33 (1928), pp. 240–5.

16 Gillian Evans, *Old Arts and New Theology* (Oxford, 1980), explains the importance of Platonic and Pythagorean mathematical influence on the teaching of theology in the Middle Ages and clarifies the relationship between the quadrivium and theological ideas (see pp. 120–35).

17 Charles Burnett, *WIST* XIV, Catalogue 22, 31, pp. 173, 177.

18 Guibert de Nogent, *Autobiographie*, ed. E.R. Labande (Paris, 1981), pp. 285, 287, 335–57.

19 Henri Pirenne, 'Northern Towns and Their Commerce', *CHME*, vol. 6, *The Victory of the Papacy* (Cambridge 1929), pp. 505–526 (511).

20 C.W. Previté-Orton, 'The Italian Cities Till c. 1200', *CHME*, vol. 5, *Contest of Empire and Papacy* (Cambridge, 1926), ch. 5, pp. 208–41 (220, 239).

21 Robert S. Lopez, 'Trade of Medieval Europe, the South', *CEH*, vol. 2, ed. M. Postan, E.E. Rich, *Trade and Industry in the Middle Ages* (Cambridge, 1952), ch. 5, pp. 257–338 (283), 2nd edn 1987, pp. 306-385.

22 I owe this information to Richard Lemay, note with letter, 11 December 1990.

23 See n. 1.

CHAPTER 4

Journey to Syria

When Adelard left Laon he advised his nephew and his other pupils to remain there and learn all they could of philosophy as it was taught in northern France. He would travel and study with the Arabs and on his return they would compare notes. *Quaestiones naturales* is the resulting essay (see Chapter 5). The period of separation lasted seven years. *Quaestiones* tells us little of Adelard's travels so one must make one's own deductions from occasional references in his works.

It is probable that Adelard made his way to Syria via southern Italy, Sicily and Greece. In *De eodem*, which he dedicated to the Bishop of Syracuse, he mentions both Greece and Salerno. In *Quaestiones* he describes being shaken by an earthquake as he crosses a bridge at Mamistra (modern Misis) near Adana on the way to Antioch. He speaks of the bridge itself and of the whole region as shaking violently with the movement of the earth, 'ipsum pontem simul etiam in totam regionem terraemotu contremuisse'.[1]

Adelard's mentioning of the earthquake is very useful in establishing a date for his journey. The earthquake is listed in J. Milne's 'Catalogue of Destructive Earthquakes', *British Association for the Advancement of Science Report* 1911, among those 'of an intensity to destroy towns and devastate districts'. It took place in 1114 and affected Anatolia. Great damage was done to Antioch, which is one hundred miles from Misis, and as far away as Edessa. The Franks were under serious threat from forces being raised against them by Sultan Mohammed. Roger of Salerno was Prince of Antioch and personally supervised repairs to the fortifications.[2]

The First Crusade had by this time succeeded in retaking Antioch and in establishing the Kingdom of Jerusalem and the counties of Tripoli and Edessa. Europeans were in control of a substantial area although they were still threatened by the forces they had displaced. Their political ambitions were such that they ignored promises to the Emperor Alexius concerning the return of territories like Antioch which had been part of the Byzantine Empire.

In theory one would have expected close collaboration among the leaders such as the Prince of Antioch and the Counts of Edessa and Tripoli and the King of Jerusalem; but all were rulers in their own right with tenuous feudal links and strong rivalries. The Kingdom of Jerusalem was fortunate in Baldwin I who reigned from 1100 to 1118. He succeeded his brother Godfrey who had founded the kingdom and his appointment was confirmed by the barons. Along the coast he was aided by Genoese and Venetians and reduced the ports of Arsuf-Caesarea, Acre, Sidon and Beirut. On the east he carried his arms beyond Jordan and built the fortress of Montreal. In times of crisis he was powerful enough to persuade the other rulers to work together.

Since it is not known whether Adelard was a recent arrival in 1114 or nearing the end of his seven years of travel, it is difficult to say which of many military incidents took place while he was there. He probably reached Cilicia and Antioch when the principality was still being ruled by Tancred, who had succeeded his Norman uncle Bohemond, the elder son of Robert Guiscard. Bohemond's uncle ruled Norman Sicily. By 1109 Tancred was one of the most powerful of the Crusading leaders, in control of territory which stretched from Tarsus and Misis to Latakia. In Antioch he was supported by the Genoese, who had rights granted by a charter, a church and thirty houses, and the Pisans who had extra-territorial rights for trading purposes.

Tancred's rival and neighbour along the coast south of Latakia had been Raymond of Toulouse who had become Count of Tripoli without taking the city itself, although he had built a great castle called Mount Pilgrim. After Raymond's death and the departure of his widow and infant son for France, Raymond's second heir, Bertrand, arrived to claim his inheritance. He was opposed by William Jordan, Raymond's regent, with Tancred's support. King Baldwin had sufficient power to persuade the three men to come to an agreement and then work together to seize the city of Tripoli, which was still occupied by a highly literate Muslim sect, the Banū Ammar.[3]

Bertrand promised a safe conduct to their governor and his forces which was honoured, but the Genoese who were assisting Bertrand sacked and burned a great college library and seized private libraries as well. The seige was described by the author of the Damascus Chronicle:

> The Franks pressed their attack upon the city and delivered an assault from their towers and captured it by the sword on Monday 11th Dhu'l-Hijja (12th July) of this year AH 502 (1109). They plundered all that was in it, took the men captive and enslaved the women and children; the quantities of merchandise and storehouses and the books of its college and in the libraries of private owners, exceed all computation ...[4]

The Arabic manuscripts which fell into Genoese hands at this time would have been exactly the sort of material in which Adelard was interested. Although it has always been considered that the scholarly works were destroyed, the possibility that some books reached merchants in Antioch should be considered. One would expect some of the Arabic books seized at Tripoli to be taken over by Adelard and his colleagues. After these events a third of Tripoli went to the Genoese and the rest to Bertrand. Tancred obtained no territorial advantage but continued to consolidate his previous conquests.

The period from 1109 to 1112 when Tancred died was one of shifting fortunes and minor battles between opposing forces. In 1111 two large armies, one of Crusaders and one of Muslim contingents, manœuvred at close quarters but did not come to a pitched battle. In December 1112 Tancred died and was succeeded by Roger of Salerno.[5]

Tarsus and Misis, places which Adelard records as having visited, had had important Greek Christian communities before the Turkish occupation. Antioch had been a centre for trade between the Greeks and the Muslims. In Antioch, after its recapture and the establishment of the Norman principality, the relations between the Orthodox Church and the Latin were strained but not so much as in Jerusalem. The monasteries were restored but Latin bishops were quickly appointed and consecrated in Jerusalem rather than in Constantinople. There were three large Benedictine abbeys and these and other religious houses were very wealthy.[6] It is possible that Adelard may have sought hospitality from them. The attitude of men like Tancred and Roger could have been favourable to Adelard's work if they had the same interest in Arabic science as that shown by the Normans in Sicily, to whom they were related through Bohemond. Antioch was a place where Arabic manuscripts could be translated into Latin. This was shown by the work of Stephen of Pisa when he translated medical works.[7] Adelard speaks of Arab masters and mentions an old man of Tarsus who explained methods of dissection for the purpose of studying anatomy.[8] Beyond that no individual appears as his mentor; nor does he refer to when and how he learned Arabic.

The general impression must be that Syria at this time was not an intellectually oriented community, and Adelard would have experienced great difficulty in acquiring what he wanted or even managing to hold conversations with Arab scholars. This raises the question of how long he stayed in the East and what places he might have visited in addition to those he himself mentions.

A fragment purporting to describe an experiment conducted in Jerusalem in 1115 has occasionally been attributed to Adelard. This implies that he reached the Holy City. Scholars now ascribe the description of the experiment to *Liber Nimrod*.[9] It should be pointed out, however, that the result described could not have occurred in the latitude of Jerusalem but only in some place in the latitude of Syene in Egypt.[10]

There is no indication that Adelard in fact reached Jerusalem, but no reason why he could not have done so along the route from Misis to Antioch and onwards. If he remained in the East after the incident on the bridge in 1114, he would certainly have been affected by some of the events which took place during this period. In 1115 the threat to the Crusaders from Sultan Mohammed intensified. Roger arranged an alliance with Toghetkin in Damascus (who was himself allied with Ilghazi, the Ortoquid of Aleppo). Nevertheless, in September Roger faced a force of 5,000 gathered against him. A stroke of good fortune enabled him to mount a surprise attack and defeat the Sultan's army, which greatly enhanced his reputation.[11]

After 1115 there was a brief period of equilibrium and it is just possible that Adelard took advantage of it. Arab scholarship had continued to flourish despite military activity. Omar Khayyám, later known for his *Rubaiyat*, was in his own day more famous as an astronomer and a mathematician. He had devised a new calendar for Malik Shah, the Seljuk ruler whose conquests were a primary reason for the Byzantine appeal to the Pope for assistance. Omar based his calendar on up-to-date astronomic observation. He had fallen into disfavour after Malik Shah's death and by the time of the Crusades had moved to Marw where at Sanjar's court a new *zij* was being produced. Al-Khazini was responsible for this – the tables were the Sindjari tables. Al-Khazini had a Greek slave who had studied geometry and one of his colleagues was preparing a commentary on Euclid. Omar Khayyám was still alive when Adelard was in the East but it is doubtful that Adelard reached Marw or even Baghdad where al-Badi al-Asturlabi, a subsequently famous astronomer who worked at the court of al-Mahmūd, had gone from Isfahan.[12]

Observatories in Islam were not like the institutions of today; they were sites or buildings where instruments could be readily employed and results recorded. The great Caliph al-Ma'mūn had built one in Baghdad in the ninth century which replaced an earlier observatory near Damascus. Nevertheless, al-Ma'mūn also had instruments constructed to be used near Damascus and various experiments took place there. The site of this observatory was near Dayr Murrän Monastery on Mount Quaysiyan. The instruments were an armillary sphere, a mural quadrant made of marble with a radius of 5 m (16 ft 5 in), and a separate accessory part sliding over its arc. There was also a gnomon 5 m (16 ft 5 in) high. The azimuthal quadrant was taken to Maragha in the thirteenth century, so it would have been still *in situ* in Adelard's time. It had been described by al-Bīrūnī.[13] Damascus is one place where Adelard might have been able to see for himself the methods of Arab astronomers. It was fairly close to the territory held by Crusading forces.

The most important of the manuscripts which Adelard acquired during his travels for future translation into Latin were the thirteen books of Euclid's *Elements* which had first been translated into Arabic from Greek three centuries earlier (see Chapter 7). The Euclid, as well as other manuscripts in which Adelard became interested, had their origins in the House of Learning set up in Baghdad by Caliph al-Ma'mūn in the ninth century. It was al-Ma'mūn who first commissioned the work of al-Khwārizmī, the mathematician and astronomer, and brought together the scholars who translated the *Almagest* of Ptolemy. He set up investigations into the value of the obliquity of the ecliptic and made an attempt to discover the actual measurement of Ptolemy's stade. Not long afterwards the astrology of Abū Ma'shar and the theories of Thābit b. Qurra were produced. Under his aegis both Euclid and Aristotle were translated.[14] It is reasonable to assume, therefore, that many of these works would exist in Adelard's day and be widely dispersed. In fact the *Zij* (astronomical tables) of al- Khwārizmī had been adapted for use at Cordova, hence the suggestion that Adelard might have found his Arabic material in Spain.[15]

It is not actually known where Adelard procured the documents on which he later worked. There is no question, however, but that the area where he says he had been was one where the Crusaders and their followers were exposed to new ideas which had had an important influence on them. Of particular interest were the style and construction methods employed by the Seljuk Turks. The architectural historian Dr John H. Harvey has commented on the fact that after the earthquake in 1114 the bridge at Misis was likely to have been repaired using pointed arches to replace the rounded ones of an earlier bridge constructed in the time of Justinian (see Plate 6). It was in Anatolia that skilful masons were using techniques that were subsequently employed by the Crusaders in their own buildings, and in some cases the workmen accompanied their new masters when they returned to Europe.[16]

Could Adelard have seen the bridge being repaired and observed the techniques? Of course we do not know, but the new arches were similar to those of the bridge at Diyarbakir across the Tigris, near to a mosque where the Seljuk Turks (in 1117–25) used arches of the type soon to be familiar in the West. Adelard does not give any indication that he is interested in architectural design as such but he shows considerable interest in proportion and measurement. It is certainly possible that he could have passed on information about what he had seen to those interested at home. It can, however, only be a coincidence that Queen Matilda commissioned the first stone-built bridge in England, at Stratford-le-Bow, before 1117, and that it was of a type never before seen in England. The movement of ideas this reflects had its beginnings before the time of Adelard's journey.[17]

A recent theory, completely at odds with this one, suggests that Adelard's eastern journey was fictional, invented to provide cover for his own ideas. Subscribers to this believe that *De eodem* and *Quaestiones* were produced in quick succession, the latter being based largely on information gathered in Sicily *c.* 1112. This implies that the bridge incident was invented or based on hearsay evidence alone.[18]

I am confident from the way Adelard refers to the event that it really happened to him. My point of view has been strengthened by the fact that John of Seville describes the early translator of part of *Liber prestigiorum Thebidis* (Adelard) as *fatuus antiochenus.*[19] I do not rule out the possibility that Adelard spent a much larger part of his seven-year stint studying Arab science in Sicily. There he could have enjoyed the hospitality of the Bishop of Syracuse. Moreover, in both Sicily and Spain there was a genuine interest in scientific research which is not apparent in the Crusading kingdom.

Opportunities for Adelard to visit places not under the control of the Crusading forces would have been extremely limited. The truce with Damascus was shortlived. Roger's increasing power worried his temporary allies. Toghetkin and Ilghazi joined forces to oppose him again. In 1116 the Turks took Roger by surprise and defeated him. In 1119 Roger failed to wait for promised support from Baldwin II and marched out to do battle near Aleppo. On this occasion at the 'Field of Blood' he lost most of his army and his own life.[20] By this time Adelard was back in England. He had probably returned by 1116.

In connection with Adelard's journeys in pursuit of scholarly material some mention must be made of the work entitled *Mappae clavicula.* One text of this 'compendium' of ancient and medieval recipes on varied subjects was attributed to Adelard in the table of contents of a thirteenth-century Royal collection, but it went missing.[21] It is thought to have been the manuscript of the work discovered in the nineteenth century by Sir Thomas Phillipps who knew nothing of the association with Adelard but deduced that the work had been written by an Englishman since there were several Anglo-Saxon terms in the course of the text.[22] The compilation itself is of the type which transmitted information across the centuries. The original *Mappae clavicula*, which became the core of later documents, was probably Greek, written down in Egypt in the fourth century, and intended as an explanation of the teachings of Hermes Trismegistus. The title, usually translated as 'A Little Key to Drawing (or Painting)', presumably was chosen to disguise the alchemical nature of the contents, and the fact that it dealt with magic and occult science.[23]

The core information in *Mappae clavicula* was a collection of recipes for the refining of gold, ersatz gold, silver and other metals; then the gilding and working in gold, silver and other metals; further dyes, other minerals and their uses, dyes for glass and dyes for leather. Later versions contained interpolated information on methods of naval warfare used by the Saracens, and other military instructions, as well as information on the specific weight of gold and silver, extracts from Palladius, and descriptions of Moorish Arabic methods of building with oil and pitch. A librarian's catalogue for the year 821–2 at the monastery of Reichenau lists a manuscript entitled *Mappae clavicula de efficiendo auro* – 'A Little Key to Drawing: Concerning the Making of Gold' – which no longer exists. Another ninth-century manuscript from northern France called Sélestat has survived. It is now considered that the title *Mappae clavicula* belonged to the lost Reichenau manuscript which was based on the original Greek text, and its contents can be deduced within the Sélestat and Phillipps versions, although in doubtful form with many additions. What is interesting is that the core material does not come from Latin sources. Where architectural information is concerned there are chapters which bear no relation to Vitruvius. Those which do are thought to have reached Reichenau by way of Charlemagne's court via Alcuin. Information on the chemical constituents of pigments which circulated in the eleventh century made its way into the Phillipps MS which is now in the Corning Museum of Glass in New York State. These recipes come at the beginning of the manuscript and are followed by the Prologue to the text which derives from the original Greek.[24] A good deal of the twelfth-century 'update' concerns matters which would have been of interest to Adelard. That is not necessarily proof that he was responsible for incorporating it.

The additions to the Phillipps MS which are of special interest in connection with Adelard's travels are those concerned with the measurement of heights by using an *orthagonium*, an instrument constructed from a right-angled triangle, Chapter 213; also Chapters 195–203, which have Arabic terms; Chapter 212 on making alcohol; and two recipes for making sugar candy from sugar-cane, Chapters 285 and 286. There are tables of runes which follow the 'Anglian' system, also a table of weights similar to that in *Regule abaci* (see Figure 3), and a multiplication table set out in a square. These are all consistent with Adelard's known interests and the fact that he had travelled in Syria. It is interesting that the ingredients for alcohol are given in code. The recipes for sugar candy or toffee are amusing because sugar cane would have been unfamiliar to a northern European. Whoever contributed these had a sweet tooth. The two English words *gatetriu* (goat-

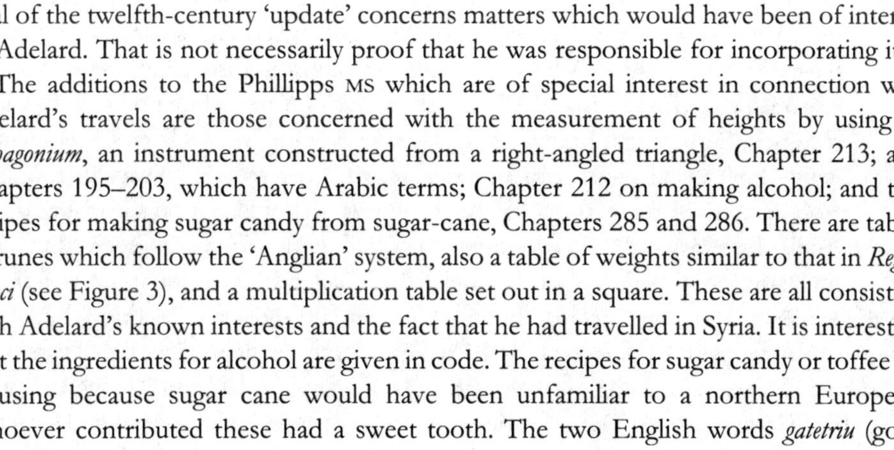

As											Uncia
12	11/12	10/12	9/12	8/12	7/12	6/12	5/12	4/12	3/12	2/12	1/12

1 Pound 1 Ounce

3 Symbols for duodecimal fractions based on the table of weights in the *Mappae clavicula*, Phillipps-Corning MS.

tree or *caprifolium*) and *greningpert* prove merely that Chapters 190 and 191 were added by an Englishman. Greningpert or Greenwort must have been 'greenweed', Dyers Greenwood, Genista tinctoria or broom. The leaves, flowers and twigs make a yellow dye, used for colouring fabrics. When mixed with woad it makes a green dye. Green was a colour in which Adelard took a particular interest.

Careful examination of these passages and comparison with other known Adelard texts has revealed that positive identification of Adelard as the author of any of the interpolated sections cannot be justified categorically; there are differences which raise doubts. But there is still support for his having been acquainted with its contents, parts of which would have been useful to him, for example in relation to his work on Thābit b. Qurra.[25]

The view that Adelard had a closer connection with the *Mappa* is still supported by some scholars, among them Christopher Hohler, formerly of the Courtauld Institute:[26]

> The particular interest of the Phillipps-Corning MS – the one 'associated' with Adelard – is that it *does* contain the first *Mappae clavicula*. It is the most recent – and the most complete – copy to survive, it is some two centuries later than the next most recent (the Schlettstadt/Sélestat MS) which is itself an epitome of the same – certainly German – redaction, and its source must have been dug out by some industrious and unusual scholar at a pretty 'late' date. What we have is a fair copy of an interpolated text of this with the kind of additions before and after all copies have; and among the *interpolations* in the model of the MS we have were the recipes with the English and Arabic words. As we have a text associating Adelard with *Mappae clavicula*, it is hardly even reasonable to dispute that the industrious and unusual scholar who found and recopied the best text known of *Mappae clavicula* and interpolated a few recipes with Arabic and English words, was Adelard. If he wasn't, he was his double. And, in view of the extreme rarity of the text, it would not be surprising if Adelard's copy of the anonymous work, with all the additions received in transcription, was described as Adelard's although he had not composed it.

On the evidence available Adelard obtained the German redaction and not a redaction otherwise known from Italy. Even if Adelard were merely acquainted with this version of the *Mappae clavicula*, he would have been struck by paragraphs 102 and 103 which appear also in the Sélestat version. They describe how to lay foundations for a vaulted structure and for building in water. At first sight Chapter 101, as transcribed and clearly shown in the Phillipps MS, appears to indicate a method of building bridges. It begins 'Dispositio fabrice de pontibus . . .'. The whole passage makes better sense if one follows, as do the modern editors, the Sélestat manuscript and translates it as an instruction on the correct depth of foundations for all structures with arches, not exclusively bridges. *De pontibus* is a mistranscription of *deponibus*. The important point is that if vaulting or arches are involved the foundation must be as deep as the height of the wall without the arch. The chapter which comes next deals explicitly with building in water and describes a sort of coffer dam. These and other chapters on building methods indicate the type of architectural problems with which contributing authors have been concerned.[27]

Harvey has made the point that no one has really fathomed the 'secrets of medieval masons' and that one of Gothic architecture's particular features is that the use of proportion derives from a source totally divorced from Vitruvius. He believes that the architects and builders who kept the 'secret' and handed it down made their initial experiments among the Saracens. The availability of Euclid's geometry as a scientific basis for their practical achievements was also important as it was to be later in Europe.[28]

Very little was written down about these matters and under deliberate concealment in some cases. But Adelard was not shy about recording information or using that which others had written down. He does not claim for himself an association with the *Mappae clavicula* but neither did he do so for some of his translations, the *Centiloquium*, for example. He was fortunate that his grounding in geometry and his interest in proportion were keen enough for him to recognise the importance that a translation of the complete Euclid geometry would have. Yet he did not undertake this at once: his first work on his return was *Quaestiones naturales*. There is no question about Adelard's authorship involved here.

Notes

1 *De eodem*, pp. 3, 33; *Quaestiones*, Müller, Introduction, p. 1 (Plan to make journey), ch. 50, pp. 49, 50 (Earthquake), comment on Adelard's journey to the East and possible dates, pp. 72–7. Müller appears not to know that the earthquake could be positively dated and suggests Adelard's journey might have ended as early as 1111 or as late as 1116. The earthquake was exceptionally violent, affecting a wide region. Harvey draws attention to J. Milne's catalogue in the *British Association for the Advancement of Science Report for 1911* (Portsmouth, 1912), p. 665. See also p. 46 below for Adelard's description.

2 Sir Steven Runciman, *A History of the Crusades*, 2 vols, vol. 2, *The Kingdom of Jerusalem and the Frankish East 1100–1181* (Cambridge, 1952), p. 130. Claude Cahen, *La Syrie du nord à l'époque des Croisades et la principauté franque d'Antioche* (Paris, 1940), p. 271.

3 Runciman, vol. 1, p. 112; Runciman, vol. 2, pp. 32–126, esp. pp. 50, 51 (Treaty of Devol), 68, 69 (Conquest of Tripoli).

4 Ibn al Qualanisi, *Continuation of the Chronicle of Damascus, The Damascus Chronicle of the Crusades*, sel. and trans. into English H.A.R. Gibb (London, 1932), p. 89.

5 Runciman, vol. 2, pp. 115–26; Robert Lawrence Nicholson, *Tancred, A Study of His Career and Work in their Relation to the First Crusade and the Establishment of the Latin States in Syria and Palestine* (University of Chicago Libraries, private edn 1940), *passim*.

6 Cahen, pp. 308, 309, 323, 324, 334.

7 David Knowles, *The Evolution of Medieval Thought* (London, 1962), p. 186.

8 *Quaestiones*, Müller, p. 21.

9 *WIST* XIV, p. 4, Catalogue 62, p. 183, identified by Peter Dronke who was kind enough to send me the following information:

> 'The passage about Mount Moriah and the centre of the world (Haskins, *Studies* pp. 31–32, 'Mons Amorreorum [better than as 2 words] … quod inquirebam') is part of the (tenth, or perhaps ninth-century) *Liber Nimrod*: it occurs (with some variation of wording) in all three complete MSS of the work: Vat. Pal. lat. 1417 (v. XI) f. 15r.; Venice Marc. lat. VIII 22 (v. XII3/4), f. 26 r.-v.; Paris B.N. lat. 14754 (XII3/4), f. 224 r. The work is still unpublished, but apart from Haskins – who clearly had not read the *Liber Nimrod* right through! – there is a good recent discussion by S.J. Livesey & R.H. Rouse, "Nimrod the Astronomer", *Traditio* 37 (1981) pp. 203–66' (letter to Louise Cochrane 1 October 1984).

It is possible that Adelard was acquainted with *Liber Nimrod*. Texts are traced in the above-mentioned article. Nimrod, other than from Genesis 10:8, was the principal character in a mythological astronomical handbook possibly known to French clerks in the English civil service. An Anglo-Norman computistical text, *Li Cumpoz*, written by Philippe de Thaon, refers to Nimrod as one of his authorities. The purpose of this work was to provide instructions for establishing ecclesiastical dates. It was written for Philippe's uncle, Honfroi, chaplain to Eudo, Henry I's *dapifer*. Philippe later dedicated a *Bestiary* to Henry I's second queen, Adeliza. See M.D. Legge, *Anglo-Norman Literature and its Background* (Oxford, 1963), pp. 18–26; C.H. Haskins, *Studies in the History of Mediaeval Science* (London, 1927, repub. 1960), pp. 336–45.

10 The experiment was an interesting one in which the author claimed to have suspended a circular plate at noon at the time of the summer solstice on Mount Moriah, in Jerusalem, and that there was a completely circular shadow. The late Professor Forbes of Edinburgh University analysed the experiment for me. The result would have been an elliptical shadow at the latitude of Jerusalem. The experiment was intended to reinforce the belief that Jerusalem was the earth's centre, a view commonly held at this time. It would have been unlikely to be held by Adelard. See below, ch. 8, p. 75.

11 Runciman, vol. 2, p. 132.

12 George Sarton, *Introduction to the History of Science* (Baltimore, 1931), vol. 2, part 1, pp. 122, 167.

13 A. Sayili, *The Observatory in Islam*, Publications of the Turkish Historical Society, Series 7, 38 (Ankara, 1960), pp. 69–81, 96, 130, 160, 175.

14 J.L.E. Dreyer, *History of the Planetary Systems from Thales to Kepler* (Cambridge, 1906), p. 244; M. Rekaya, 'al Ma'mūn', *EoI*, vol. 6, (1990), pp. 331a-39b.

15 Haskins 1927, pp. 20–42 (34).

16 John H. Harvey, 'The Origins of Gothic Architecture', *Antiquaries Journal* 48 (1968), pp. 91–4; see also Christopher Brooke, *The Twelfth Century Renaissance* (London, 1969), pp. 102, 103.

17 John H. Harvey, 'Geometry and Gothic Design', *Transactions of the Ancient Monuments Society* 30 (1986), pp. 47–56, p. 48, p. 56, n. 25.

18 Brian Lawn, *The Salernitan Questions* (Oxford, 1963), p. 30. Gibson, *WIST* XIV, p.7, n., p.10, n.25, p.11.

19 Richard Lemay, 'The True Place of Astrology in Medieval Science and Philosophy: Towards a Definition', in Patrick Curry (ed.), *Astrology, Science and Society* (Woodbridge, Suffolk and Wolfeboro, New Hampshire, 1987), p. 70.

20 Runciman, vol. 2, pp. 148, 149.

21 Haskins 1927, p. 30. *WIST* XIV, Catalogue 25, p. 174, 'A work on making pigments and other artifacts; attributed to Adelard in the list of contents of *London, BL, Royal 15.C.IV*, which no longer includes the work'. See also Catalogue 64, p. 183. *Mappae clavicula* has been translated 'A Little Key to Drawing' but could be translated in other ways. The full title of one manuscript is *Libellus dictus mappae clavicula* ('A Little Book Said to be a Little Key to *Mappa*'). Since *Mappa mundi* ('Map of the World') became *Mappemonde* and *mappa* was the shortened version, 'A Little Key to the World' is a further suggestion. The original meaning of *mappa* was 'tablecloth' or 'napkin', then 'covering', and later 'plan' and 'book' (from linen cloth). It possibly was used in monasteries as the title of the cover for a collection of folios, i.e. portfolio. So another possibility would be 'A Little Key to the Portfolio concerning...'.

22 Sir Thomas Phillipps and Albert Way (eds), 'Mappae Clavicula, a Treatise on the Preparation of Pigments during the Middle Ages', *Archaeologia* 32 (1847), pp. 183–244.

23 Bernard Bischoff, 'Die Überlieferung der technischen Literatur', *Mittelalterische Studien* 3 (1981), pp. 277–97.

24 C.S. Smith and J.G. Hawthorne, (eds), 'Mappae Clavicula, A Little Key to the World of Medieval Techniques', *TAPS* 64, part 4 (Philadelphia, 1974).

25 Charles Burnett and Louise Cochrane, 'Adelard and the *Mappae clavicula*', *WIST* XIV, pp. 29–32; Smith and Hawthorne, chs 195–203, pp. 57, 58; chs 212, 213, p. 59; chs 285, 286, p. 71; Fraction tables p. 73; chs 190, 191, p. 54. Greenweed is discussed by George Usher, *A Dictionary of Plants Used by Man* (London, 1974) pp. 271, 272. For Hermes Trismegistus, see ch. 9, below.

26 Christopher Hohler, formerly of the Courtauld Institute, letter to Warburg Institute, 18 May 1988. Permission to quote has been received.

27 Smith and Hawthorne, p. 42.

28 Harvey 1986, pp. 44, 48.

CHAPTER 5

Return to England – *Quaestiones naturales*

On his return from abroad Adelard's first literary effort was the writing of *Quaestiones naturales*. He speaks of his almost immediate desire to discover the manners and customs of his own country but that he has learnt that 'its chief men are violent, its magistrates wine-lovers, its judges mercenary, its patrons fickle, private men sycophants, those who make promises deceitful, friends full of jealousy, and almost all men self-seekers'.[1]

Is there justification for Adelard's comments on the state of the country at this time? Possibly. Patronage was important to Henry I's administration. The organisation of Henry's justice may have seemed less attractive to his contemporaries than it has in longer perspective.[2] Certainly the early problems of the reign were now resolved, but Henry rarely deferred to the opinions of his bishops or men of high standing. The rules established by his Curia did make large concessions to feudalism, and feudal rights were tolerated unless expressly abolished by royal authority. This meant that archbishops, bishops, earls and other officials had the power to administer justice in lands under their jurisdiction. The charters for Bath show that Henry conferred wide powers: all the rights which the Crown had previously possessed there had been extended to Bishop John. But every tenant must regard the king as his chief lord. An incident involving Bishop John illustrates that tensions could arise.

Henry's son, Prince William, had a friend who accused the Bishop of Bath of unjustly refusing to enfeoff him with (grant him possession of) an estate which he had purchased from a tenant of the bishop's. William, as his father's deputy, issued a writ directing the bishop to give seisin to the claimant. The bishop took counsel with his tenants, decided that neither the king's son nor the king himself had a right to ask for more than was just, and refused to obey the writ on the ground that the alleged claim was invalid by the custom of his court.

The tragedy of the White Ship, the shipwreck which caused the death in 1120 of Henry's heir, William, prevented matters from going further. The bishop's protest shows that the custom of appealing to the law against the Crown was already a familiar one on English soil.[3] Nevertheless, it would require great strength of character, and this kind of quarrel over property would rouse strong attitudes among leading citizens of Bath.

Adelard may have been made aware of local bitterness on his return. Bishop John had originally acquired his property in Bath from William II. His relations with Henry had been good but a quarrel with Henry's son would make things difficult and perhaps affect Adelard as well.

This perhaps explains why Adelard dedicates *Quaestiones naturales* to Richard, Bishop of Bayeux, rather than to John of Bath. Richard was a member of an Anglo-Norman ecclesiastical family, son of Samson, Bishop of Worcester, and nephew of Thomas, Archbishop of York. Like them Richard probably reached the episcopate through the usual agency of a royal chaplaincy. He was appointed in 1107. In dedicating *Quaestiones* to him Adelard declared that nothing in the liberal arts had been so well treated that it would not have 'bloomed more richly' had it been composed by Richard.[4]

It is not known at what date Adelard became associated with the court of Henry I. His name is often mentioned with that of Petrus Alfonsi, considered to have been one of King Henry's physicians, a converted Jew from Spain who is thought to have been in England from 1112 to 1120. Both men were important in the transmission of Arabic science to the West and had links with the court of Henry I. Both worked on the *Zij* of al-Khwārizmī, although whether this was separately or not cannot be proved.[5] If there were a collaboration, however, there was no necessity that it begin in 1112 as long as Adelard was back in the court surroundings of Henry I well before 1120 in time to establish a working partnership. So this would not preclude a trip to Syria and possibly further east. The choice of 1116 as an appropriate date for his return seems reasonable and allows seven years for travel after leaving Laon.

Quaestiones marks a transition in Adelard's thinking from the philosophy in *De eodem* with its emphasis on generalisations to his increased interest in natural philosophy and the application of 'reason' to scientific method in arriving at his conclusions. Adelard's nephew again has a role as a literary device. He asks the questions which challenge Adelard who gives the replies. Adelard at first implies that his answers are those which his Arab teachers would give, but later drops this pretence to give forthright opinions of his own.[6]

Attention has been drawn to the fact that Adelard could have written *Quaestiones* solely on the basis of what he had learned in Salerno. And despite a suggestion that Adelard introduced the work of Constantine Africanus to Europe, it has also been said that there is no indication of acquaintance with Constantine's writing in Adelard's text. Some Aristotelian ideas occur but nothing to show that Adelard has seen Aristotle's *Libri naturales* or done more than listen to the philosophic notions of peripatetic scholars.[7] Still, ideas from Salerno would be as new to England as those from Antioch, and it is best probably to view this work as a summary of the conclusions Adelard has reached since he completed *De eodem*. Whether the journey was fact or fiction, copies of the manuscript and later the book entitled *Quaestiones Naturales* exist, and Adelard is the author.

There would have been great interest in medicine and natural science in Bath where Bishop John was actively engaged in exploiting the mineral hot springs. A further point can also be made: Arabic natural philosophy was imbued with ideas from Plato and Aristotle as well as Hippocrates and Galen. The framework of Arabic science was almost exclusively Greek, though tempered with Indian and Persian ideas. The Arabs did not look on translation as a purely mechanical or menial task. Al-Kindi had stressed the fact that the mere reading of ancient texts could become an obstacle to clear thought unless used as a call to consult nature herself and solve apparent contradictions in that way.[8] This is exactly what Adelard does. He is thinking independently about his new sources of knowledge. That he had begun to do this in Salerno does not mean he would not continue to do so if he travelled to Antioch, and it is understandably difficult to trace sources for every point that he makes.

Adelard attributes to Arab teachers an attitude of mind which was already quite familiar to members of the Chartres school:

> I with reason for my guide have learned one thing from my Arab teachers – you something different – dazzled by the outward show of authority you wear a headstall, for what else would we call authority but a headstall? Just as brute animals are led by headstalls where one pleases so many of you are led into danger by the authority of writers ... [9]

Berengar of Tours in the eleventh century had been one of the first to preach that Reason and not Authority was mistress. The emphasis on reason was well to the fore in *De eodem*. It was important in the thinking of theologians such as Archbishop Anselm. Adelard should not perhaps have implied that the point of view was uniquely 'Arab'.[10] He certainly emphasises that learning from previous authorities should bring new ideas and not merely perpetuate what men have thought before. He remarks that his contemporaries feel that nothing found by 'moderns' should be received and this is the sort of thinking which prevents progress. The approach encouraged by Adelard is said to have completely changed the state of natural philosophy, but clearly he was apprehensive about the views he was going to express.[11]

When Adelard speaks first of his desire to have the opinions he gives considered to be those of his Arab teachers rather than his own, he is probably being careful. He had just returned. Support for mathematics and astronomy *qua* astrology was not universal among churchmen. Adelard's contemporary, the historian William of Malmesbury, described Gerbert's mathematics as 'dangerous Saracen magic'.[12] William was also disturbed to find that Bishop Gerard, Robert of Hereford's successor and later Archbishop of York, was interested in astrology. William was a protégé of Queen Matilda, a likely sponsor for Adelard at Henry's court, and Matilda's death on 1 May 1118 could also have affected his position.

Adelard gains confidence as he proceeds, however, and later in the text he confesses that he is not sure whether he is expressing his own ideas or those of one of his teachers.[13] He is not afraid toward the end of his work to challenge those who disagree with him. His sense of humour has also been restored. When his nephew asks him how the earth remains suspended in mid-air, he replies that it would be inexpedient if this were not so.[14]

The theory of the universe in *Quaestiones* is based, as were the ideas in his earlier work, on the *Timaeus*. Adelard again assumes that his readers are well acquainted with this and begins his first chapter on plants with a discussion of the four elements, earth, water, air and fire. One should recall that the belief that all nature was the result of various combinations of the four elements and the qualities deriving from them was also shared by all Arabic scientists. One of the most important of Plato's ideas which had become acceptable to Christians was that the universe had been created by God from pre-existent matter and independent of time. Subordinate deities were then assigned the task of creating mortal souls, in part from immortal incorporeal soul and in part from other matter. *Quaestiones* provides a good illustration of the dilemma which students of philosophy and theology face when theories of how the universe functions conflict with theories about God the Creator. Adelard takes his discussion up to this point and defers consideration of God until another time. In the process it is apparent that he has been introduced to ideas which he did not have when he wrote *De eodem*. He does not, however, seem to have become acquainted with Avicenna's version of Aristotle or the work of al-Farābi or al-Ghazali. These philosophers in Islam also faced problems of interpretation when logic and religion came into conflict. Muslims believe, as do Christians, that God exists outside the universe and outside time. Aristotle's belief that the universe was eternal and that the Prime Mover was the outermost sphere beyond the *aplanos* ('firmament') could not be reconciled with Christian and Muslim religious beliefs as Plato's concept of God could be. Moreover, Aristotle denied the resurrection of the body. Avicenna's version of Aristotle was challenged in Islam before the Church challenged Aristotelianism in Europe, despite the neo-Platonism incorporated in it.[15]

The problem for Muslims was the belief that the universe was created *ex nihilo* – nothing was eternal except God and his attributes. Avicenna (died AD 1047), in translating Aristotle, followed Plato in stating that God acted more as craftsman on pre-existent matter. Al-Ghazali (died AD 1111) argued in favour of creation *ex nihilo* and belief in resurrection of the dead. His views became widely accepted in the Arab world and when translated into Latin were employed in Christian treatises, although his objectivity was misconstrued as agreement with Avicenna.

In the twelfth century Averroes (died AD 1198) claimed that Avicenna and al-Ghazali had espoused their view in order to conform with prevailing opinion, that they did not grasp the true meaning of the Aristotelian tradition with the existence of a world soul and no resurrection of the body; moreover, that God acted within eternity and not outside it.[16]

Quaestiones was written before the complete Aristotelian corpus reached the West but when many Aristotelian ideas, unknown heretofore, were beginning to be discussed. Just as *De eodem* reflects the very considerable importance of the argument about universals, *Quaestiones* reveals the thought processes which will inevitably force concentration on problems raised by a more accurate translation of Aristotle. This work of Adelard's is interpreted as advocacy of Aristotelian thought and experimental method. It demonstrates the intellectual adjustment and assimilation which Adelard engaged in before he began the systematic translation of mathematical and astronomical/astrological texts. It should be underlined that Adelard specifically limited the scope of his work to natural science. He was not writing theology.[17] This was the important difference and emphasised by David Knowles as well as Alistair Crombie:

> Down to the end of the twelfth century the predominant theological interests of scholars had led them generally to treat the natural world as a kind of shadow and a symbol of divine power and providence. The context of

> Adelard's use of reason marks the first explicit assertion in the Middle Ages that recognition of divine omnipotence did not preclude the existence of proximate natural causes, and that these could be known only by independent scientific inquiry. Though relying mainly on *a priori* reasoning Adelard had some recourse to observation and experiment.[18]

Adelard makes this point in an early chapter: 'I will detract nothing from God, for whatever is is from Him .... We must listen to the very limits of human knowledge and only when this utterly breaks down should we refer things to God'. In other words the study of natural causes can be justified in its own right as separate from theology.[19]

The arguments which Adelard uses for his replies have all been carefully thought out. Over and over again, however, he returns to his theme that it is essential for men to use the powers of reason with which they have been endowed:

> Unless reason were appointed to be the chief judge, to no purpose would she have been given to us individually; it would have been enough for the writing of laws to have been entrusted to one, or at most to a few, and the rest would have been satisfied with their ordinance and authority. Further, the very people who are called authorities first gained the confidence of their inferiors only because they followed reason; and those who are ignorant of reason or neglect it justly deserve to be called blind ... hence logicians have agreed in treating the argument from authority not as necessary, but probable only.[20]

Adelard's nephew asks him why human beings do not have horns; Adelard enquires, 'Why should they?'. The nephew returns that horns would be a means of protection. Adelard's answer:

> As man is the creature dearest to the Creator it is not fitting that he should be naturally provided with weapons. He has reason by means of which he can subdue brutes...
>
> ...Man is a rational and a social animal adapted for two operations, action and deliberation, or as we call them war and peace. Daily life teaches him that in activities of war arms are required, but in time of peace truth teaches him to lay them aside and remove them from the innermost chamber of his thoughts, for the one is provoked by wrath, to the other reason gives its sweetness .... If man were provided with natural weapons, he would be unable when engaged in making treaties of peace, to lay them aside.[21]

In Chapter xxxviii the nephew asks how is it that human beings cannot walk as soon as they are born while brutes can. Adelard again returns to his thesis:

> Weakness of limb and difficulty in walking upright arise from nature's dignity in acting as agent. The limbs are in consonance with reason and nature has taken this way of developing them so that man will not fall away from reason through contemplating his own body.[22]

Having dealt with humanity Adelard and his nephew move on to discussion of cosmological questions. It is here that Adelard is asked how the globe is supported in the middle of the air and replies 'Certainly it is inexpedient that it should fall'. He then goes on to explain that everything earthly tends toward the lowest position. Since the middle point remains fixed, middle and lowest are the same, that is, the central point of the earth can be said to be the lowest point. In reply to a question about a hole straight through the earth – in what direction would a stone thrown into it fall – he states that what causes the stationary position of the earth would produce equilibrium in the stone and it would settle in the centre. He then makes a further point. All nature loves its like and shuns its contrary. The space environing the world is the home of fire and this upper space would be shunned by earth, hence a stone through its own weight seeks that for which it has a fondness and shuns what it does not like.[23]

This theory, taken from the *Timaeus*, had been enunciated earlier in *De eodem*. Adelard uses it now as a lead for the discussion of earthquakes. His nephew refers to Adelard's report on the one he had experienced on his travels:

> *Nephew*: If indeed as you explained about heavy bodies which have gained a central equilibrium for a double reason, what is it which makes them change from this causal necessity and produces instability? You will not deny that you once told me that while in the Antioch district you were crossing a bridge in a certain city, not only the bridge but the whole of the district was shaken by so violent an earthquake that you seemed to be in as great danger on land as by sea. What then is this power which is so violent that it changes the positions of things, and earth herself, that ponderous mother of weights, from the position she has taken?
>
> *Adelard*: Your anecdote is true. Just as I passed over the shaking bridge without a scratch, so I shall answer your question without damage to my consistency. In an earthquake the earth is affected particularly and not universally, and the cause of the movement I hold to be not its own quality, but the effect of its continent. By its continent I mean the air, for the air is not only diffused all round the earth but also fills its interior bowel-fashion. When therefore the outer air calls to the inner air, the latter hastens to go forth to its kin and gathering together with this object fills the caverns of the earth; and so long as it finds obstacles, shakes it violently and does not rest till it finds its course outwards. No doubt it will be obvious to you that while every such compound may be imprisoned for a time, yet at last it will strive to force its way to its original source.[24]

Shortly afterwards Adelard deals with tides. Neither his reply nor the original question suggests any connection with the moon, which is surprising since Adelard should have been acquainted with Bede's work, and in some of his other theories he has shown himself to have read Macrobius.[25]

Adelard's physiological ideas are based particularly on the interaction of the humours. Through the blood we get the right proportions in the mixture of the four chief moistures of the organism: yellow gall and black, blood and mucous, each of which has its own special seat. Each has two of the distinctive qualities: warm, cold; moist, dry. The somewhat gruesome account of the old man in Tarsus, who told Adelard that the way to study sinews was to suspend a corpse in a river until the skin and flesh fell off and veins and sinews were revealed, demonstrates Adelard's interest in anatomy.[26] When the nephew asks him which part of the brain is the seat of various functions, Adelard explains that if fancy is impaired by an injury to the front of the head, observation of this fact would help to establish a differentiation, if reason and memory were not harmed. Aristotle is given as the authority for the subdivisions of the brain, with fancy in the front, reason in the centre and memory at the back.[27]

A surprising question is one on why men go bald in front. Adelard's opinion is that vapour from the stomach reaches the front of the head first. Heat opens the pores and this prevents the hair from growing.[28]

Adelard takes his theories of sense perception from Plato and Boethius. The interpretation of how what we see is recorded in our minds is particularly vivid. Adelard dismisses the view of the Stoics that images are imprinted in our minds like writing on paper (parchment). He states very positively 'I am no Stoic, I tell you, but a citizen of Bath, and have no need to train my intellect in the mistakes of the Stoics'. He gives in this connection a detailed analysis of 'fiery breath' or 'force' generated by the brain which makes its way from the mind along nerves to the eyes and then to the body to be seen and then back at wondrous speed, and being impressed with the shape of the body it both receives and retains the shape it has received.

Adelard's nephew comments on the swiftness with which the fiery breath can travel up to the very firmament (*aplanos*) itself, that is, to the outermost part of the universe and back, and says that for a body to travel in this way is madness. Adelard's reply refers again to reason: 'The incorporeal eye of the mind is their guide, just as they see clearly the magnitude of the external continent or boundary and the almost infinite extent of the heavens and the unutterably swift movement of the visible breath, and this in both cases thanks to the use of reason.' The nephew insists that Adelard differentiate between the Stoics' impression and his own. Adelard explains:

> The Creator of this wide universe has endowed the sensible shapes of all creatures with the same pre-excellent beauty. He has bestowed it in what we call the mind. Hence it is in His power to re-embody the past, to make what is absent present and to foretell the future. Consequently He does not bring forth the treasury of sensible shapes always but only when it is necessary or when it pleases Him to do so. The visible breath, therefore, being marked by a shape unknown up to that moment, the mind displaying not that shape but a similar one, expresses the quality of its own abundance and is thence provoked into an act of judgement. Since therefore both coming and going present no difficulty to the visible fire, nor is the passing on of shapes any hindrance, let all honours and love be given to Plato, the discoverer of this divine scheme, and his followers.[29]

Despite the reference to Plato there is a strong similarity between this view and one expressed in the work entitled *De differentia spiritus et anime*, by Qustā b. Lūqā, which is a commentary on Aristotle's *De anima*. A translation of this is attributed to Adelard by John Dee, although John of Seville is considered to be the original translator.[30]

In Chapter lviii the nephew describes a visit to a witch where he and Adelard have been shown a device where a vessel of water is held upside-down but no water is released unless the servant unblocks holes in the lid. Adelard comments that if this was magic it was nature's magic. He explains the device in terms of a theory of universal continuity, the operation of a vacuum. Roger Bacon's explanation in the next century is a similar one.[31]

As Adelard reaches the climax of his book he must deal with problems of motion and it is apparent that he has given some thought to Aristotle's idea of a Prime Mover. The nephew raises the problem of what was the original force to set air in motion. Motion must be produced by something in which motion exists; therefore either there is no such thing as wind or there is universal motion. Adelard in his reply makes the somewhat obscure statement that he believes rest to be the cause of motion. He then explains:

> Whatever, however, moves, if the term is employed properly and in the sense of passivity, is certainly moved by something else; and you will admit that rest and motion have different meanings. Rest is the result of passivity, while motion leads to action, for a thing will move, and not itself be moved by something else. In form it will be active, not patient; it will be the cause of motion, not its effect. In fire, when it rises upwards, this cause or form is called lightness; in a stone, when it falls downwards, gravity; and in fire let us call it agility, for this in fire, though itself quiescent, impels its foundation to motion. Motion, therefore, as defined will be neither infinite nor will it land us in an unending search for causes. Let other people look to what they have said about motion; I for my part consider the forms of things to be the causes of passive effects; for these at times by their efficient power, which by likeness of nature calls out those of other compounds, impel the things on which they act to suffer divers effects, though they themselves undergo nothing of the sort. For this reason the original cause of all things, though in a sort it moves all things, is yet itself subject to no change; and therefore it does not follow, that because a thing produces movement, it is therefore in a state of movement.[32]

This explanation is followed by a challenge from the nephew asking if infinite movement is not set up when something moves something else, if one atom is moved, are not all moved? Adelard states that the motion of winds is orbicular, returning to where they started. Varying movements also counteract each other. The impulse of all does not follow the movement of any single one: no one movement will go on perpetually, a thing easily seen as in a ray of sunlight.[33]

In the matter of universals Adelard had observed a difference in point of view between Plato and Aristotle and produced his own explanation to reconcile the two theories. He does not attempt this here. The problem of different interpretations occurs again; then the nephew asks whether in Adelard's opinion the stars are animate or not. Adelard takes the view as does Plato that they are animate, composed of four elements and a more abundant share of what is most adapted for life and reason, fiery rather than terrestrial. He then goes on to say:

> Now we must take note of their activities in which we accept not my view but Aristotle's. If things which are in a state of motion sometimes come to rest and at other times move then they are not moved by nature, for whatever is moved by nature neither comes to rest nor changes its motion. It remains

> therefore that they are moved by force or by will. Now there is no force in nature more powerful than the turning force of the *aplanos*, but they are not moved by this, for they travel in the opposite direction. It follows then that they are moved by will and the consequence of this is plain ....
>
> Furthermore if their action determines the life or death of lower animals there is only one view we can hold about them .... We must admit that it is impossible to imagine that what produces life in others is itself devoid of life. Further it is certain that whatever observes a determinate arrangement and a fixed principle in its movement must employ reason and nothing can have a more definitive arrangement or a more absolute order than the course of the stars.
>
> ...To have knowledge without also having mind is impossible and the stars are therefore animate and possess reason and knowledge.

Adelard's mention of Aristotle's ideas is thought to have been taken from Cicero's *De natura deorum* and not from Avicenna or al-Ghazali. Lemay feels it is important to stress that Adelard is speaking of the planets as stars which have reason and knowledge.[34]

The discussion reflects Adelard's powers of observation. The star we see on the meridian at midnight is on the meridian earlier each night until we no longer see it, gradually obscured either by the light of the sun or the effect of the obliquity of the ecliptic. Adelard tells his readers that the force which causes this is the turning force of the *aplanos* and the planets move independently. For Plato the *aplanos* was part of the world soul. God was beyond, outside the universe. For Aristotle the *aplanos* was moved by the Prime Mover, the outermost sphere, within the universe not outside it, hence the importance of the question.

The literary form in *Quaestiones* is more direct than the earlier allegorical presentation in *De eodem* and the views expressed are more specific. The statement that planets are intelligent beings and move voluntarily is not one which Christians could accept. Bernard Sylvester takes a similar position in his *Cosmographia* but uses allegory to convey his meaning.[35]

The nephew's final question on whether the *aplanos* is inanimate, animate or a god produces Adelard's refusal to discuss the nature of God until another occasion.[36] He breaks off the argument when it becomes essential to consider the problem which asserted itself when Averroes undertook his own translation of Aristotle to replace that of Avicenna who had compromised on this issue.

At the end of one of Adelard's later chapters he attempts to explain thunder and lightning and refers to the relative speeds of sound and light. He proceeds to the formation of hail and the origin of thunder. He then pleads with great feeling for the investigation of causes:

> The mind imbued with wonder and a sense of unfamiliarity shudderingly from a distance contemplates effects without regard to causes and so never shakes off its perplexity. Look more closely, take circumstances in their totality, set forth causes, and then you will not be surprised at effects. Do not be one of those who prefer ignorance to a close examination.[37]

Alistair Crombie has pointed out that the development of this form of rationalisation (making use of a distinction ultimately deriving from Aristotle between experimental knowledge of a fact and rational knowledge of the reasons for or cause of the fact) was part of a general intellectual movement in the twelfth century. Not only writers like Adelard of Bath and Hugh of St Victor but also theologians like Anselm and Abelard tried to arrange their subject-matter according to this mathematical deductive method. Like good disciples of St Augustine and Plato they held that the senses were deceitful and reason alone could give truth.[38]

Despite the confident and positive affirmations throughout the book, Adelard also complains after he has talked about seeking causes:

> In matters of this sort I have found almost all men wrong-headed; and hence, when I lay down premises of this sort in talking with them, they neither accept my premises, nor listen to my explanations: they point me out with the finger and make me the subject of scandal.

This passage has given rise to the impression that Adelard perhaps met formal opposition to some of his views despite his insistence that he was not discussing God. He did not disguise his views by presenting them allegorically, and since he and other cosmologists of his generation were at the forefront of a new intellectual development, it would obviously take time for some of the theories to find acceptance. Hermann of Carinthia had similar difficulties.[39]

Adelard has developed considerable maturity in his writing since the earlier works. His personality emerges to enliven the text. In a discussion about different aspects of the colour 'green' he reveals that he is wearing a green cloak and has a ring with a green stone, presumably an emerald. The ring might have had an astrological significance related to the theories of Thābit's *Prestigiorum* which Adelard translated.[40]

Quaestiones naturales became a highly popular book, widely read throughout the Middle Ages. It was translated into Hebrew and certain alterations were introduced so that it would conform with Hebrew doctrine. Adelard's own work, an original piece of writing rather than a translation, can be considered his major contribution to twelfth-century literature.

Notes

1 *Quaestiones*, Gollancz, Preface, p. 90; *Quaestiones*, Müller, p. 1. I have quoted from the only translation available in English, Gollancz. Dr Charles Burnett is engaged on an improved translation.

2 Judith A. Green, *The Government of England under Henry I* (Cambridge, 1986), pp. 171–93, (172).

3 H.W.C. Davis, *England under the Normans and Angevins* (London, 1949), pp. 134, 135; W. Farrer, *An Outline Itinerary of King Henry the First* (Oxford, 1920), p. 83.

4 S.E. Gleason, *An Ecclesiastical Barony of the Middle Ages: The Bishopric of Bayeux 1066–1204* Harvard Historical Monographs (Cambridge, Mass., 1936), pp. 23, 24. Adelard's letter printed by E. Maarlène and V. Durand, *Thesaurus Novus Anecdotorum*, 5 vols (Paris, 1717), vol. 1, p. 291.

5 E.J. Kealey, *Medieval Medicus* (Baltimore, 1981), pp. 46, 66–80. Professor Kealey was kind enough to lend me his notes on a special study in preparation on Adelard, Walcher of Malvern and Petrus Alfonsi. See also J. Millás-Vallicrosa, 'La aportación astronómica de Pedro Alfonso', *Estudios sobre historia de la ciencia española* (Barcelona, 1949), pp. 65, 79, 105–8, 113; *Nuevos estudios sobre historia de la ciencia española* (Barcelona, 1960), pp. 87–97, 107, 197.

6 Dorothea Metlitzki, *The Matter of Araby in Medieval England* (New Haven, Conn., 1977), pp. 49–55; *Quaestiones*, Gollancz, p. 92; Müller, p. 79.

7 *Quaestiones*, Müller, pp. 87–91; Brian Lawn, *The Salernitan Questions* (Oxford, 1963), pp. 24–31, *The Prose Salernitan Questions* (London, 1979) p.390; *Quaestiones*, Gollancz, p. xv.

8 R. Arnaldez and L. Massignon, 'Arabic Science', in René Taton (ed.), *Ancient and Medieval Science, A General History of the Sciences*, 4 vols, vol. 1 (London, 1963), pp. 385–421 (pp. 392, 393). Metlitzki, p. 51, describes an early Arabic analogue containing natural questions.

9 *Quaestiones*, ch. vi, on why man must use reason with which he is endowed, Gollancz, p. 98; Müller, p. 11.

10 Gibson, *WIST* XIV, p. 10.

11 A.G. Molland, 'Medieval Ideas of Scientific Progress', *Journal of the History of Ideas* 39 (1978), pp. 561–7 (566, 567).

12 William of Malmesbury, *History of the Kings of England*, trans. Revd John Sharpe (London, 1815), p. 199. See above, ch. 1, p. 7.

13 *Quaestiones*, Müller, p. 79.

14 *Quaestiones*, ch. xlviii, on how the globe remains supported in the middle of the air, Gollancz, p. 137; Müller, p. 47. Adelard has previously referred to the problem in *De eodem* (see ch. 2, pp. 15, 19, and n. 10; also ch. 4, p. 34 and n. 9, pp. 39–40).

15 Arthur Hyman and James Walsh (eds), 'Islamic Philosophy', *Philosophy in the Middle Ages* (Indianapolis, 1974), pp. 203–59, (234, 235). Richard Sorabji discusses the philosophical problems in *Time, Creation and the Continuum* (London, 1983).

16 Soheil M. Afnan, *Avicenna, His Life and Works* (London, 1958), pp. 25–31, 126–38.

17 Knowles, 1962, pp. 132, 280; Tina Stiefel, 'Twelfth-century matter for metaphor: the material view of Plato's *Timaeus*', *British Journal for the History of Science* 17 (1984), part 2, no. 56, pp. 169–85.

18 Alistair Crombie, 'Science', in A.L. Poole (ed.), *Medieval England*, 2 vols (Oxford, 1958), vol. 2, ch. 18, pp. 571–604 (579).

19 *Quaestiones*, ch. iv, on why belief in God does not preclude use of reason to study natural science, Gollancz, p. 96; Müller, p. 8.

20 *Quaestiones*, ch. vi, on why man must use reason with which he is endowed, Gollancz, p. 99; Müller, pp. 11,12.

21 *Quaestiones*, ch. xv, on why human beings do not have horns, Gollancz, p. 106; Müller, p. 21.

22 *Quaestiones*, ch. xxxviii, on why human beings cannot walk when born, Gollancz, p. 130; Müller, pp. 41, 42.

23 *Quaestiones*, chs xlviii, xlix, on how the globe remains supported in the middle of the air, on why a stone thrown into a hole on the earth's surface comes to rest in the centre, Gollancz, pp. 137, 138; Müller, pp. 47, 48. See above, n. 14.

24 *Quaestiones*, ch. l, on earthquakes, Gollancz, p. 139; Müller, pp. 49, 50. See above, ch. 4, p. 32; nn. 1, 2, p. 39.

25 *Quaestiones*, ch. lii, on tides, Gollancz, p. 140; Müller, p. 51. Mention of Macrobius, ch. xxxv, Gollancz p. 128; Müller, p. 40.

26 *Quaestiones*, ch. xvi, on Old Man of Tarsus, Gollancz, p. 108; Müller, p. 21.

27 *Quaestiones*, ch. xviii, on brain and its divisions, Gollancz, p. 109; Müller, pp. 22,23. Also mentioned in *De eodem*, p. 32 (see above, ch. 2, p. 17 and n. 17, p. 21).

28 *Quaestiones*, ch. xx, on baldness, Gollancz, p. 111; Müller, p. 24.

29 *Quaestiones*, chs xxiii-xxx, on optics, Gollancz pp.115–24; Müller, pp. 23–36. Reference to Stoics and Adelard's claim to be a citizen of Bath, ch. xxviii, Gollancz, p. 122, Müller, p. 34. Edward Grant, *A Source Book of Medieval Science* (Cambridge, Mass., 1974), reproduces Adelard's chapters on optics.

30 Lynn Thorndike, *A History of Magic and Experimental Science*, 2 vols (New York, 1923), vol. 2, p. 33; Burnett, *WIST* XIV, Catalogue 5, p. 167, Catalogue 139, p. 194.

31 *Quaestiones*, ch. lviii, on a vacuum, Gollancz, pp. 143, 144; Müller, pp. 53, 54; George Sarton, *Introduction to the History of Science*, (Baltimore, 1931), vol 2, part 1, pp. 167–9.

32 *Quaestiones*, chs lx, lxi, on motion, Gollancz, pp 145–7; Müller, pp. 55–7.

33 John D. North, *Stars, Minds and Fate, Essays in Ancient and Medieval Cosmology* (London and Ronceverte, West Virginia, 1989), ch. 18, 'Celestial Influence', pp. 243–98 (251). In this article North notes 'that Adelard of Bath in his *Quaestiones naturales* talks as though natural motion can only be up or down and thus rejects it for the heavens. We have already seen that Cicero had the same view and it seems to me that Adelard is taking the whole relevant passage from him'. See also Theodore Silverstein, 'Adelard, Aristotle and the *De Natura Deorum*', *Classical Philology* 47–8 (1952–3), pp. 82–5.

34 *Quaestiones*, ch. lxxiv, on stars, Gollancz, pp. 158, 159; Müller, pp. 65–7; Lemay suggests that much of this argumentation is based on Abū Ma'shar, *Introductorium maius*, I,3, with Adelard's own speculation at work. Note with letter to Louise Cochrane, 11 December 1990. (See below, ch. 9, pp. 87–90.)

35 Peter Dronke, *Bernardus Sylvestris Cosmographia* (Leiden, 1978), p. 29, *Fabula* (Leiden, 1974), pp. 138, 139, n.1.

36 *Quaestiones*, ch. lxxvi, on deferring discusion of God, Gollancz, pp. 160, 161; Müller, pp. 68, 69. ('Ought we to regard the *aplanos* as an inanimate body, or a living thing, or a god?')

37 *Quaestiones*, ch. lxiv, on thunder and lightning, investigation of causes, attitudes to Adelard and his work by others, Gollancz, pp. 149, 150; Müller, pp. 58, 59.

38 Alistair Crombie, *From Augustine to Galileo*, 2 vols (London, 1951), vol. 2, pp. 3, 4.

39 Tina Stiefel, 'Science, Reason and Faith in the Twelfth Century: The Cosmologists' Attack on Tradition', *Journal of European Studies* 6, part 1, no. 21 (1976), pp. 1–16. See also Hermann of Carinthia, *Astronomia*, and his Preface to *Almagest*, Richard Lemay, note with letter, 11 December 1990.

40 *Quaestiones*, ch. ii, 'Green cloak and emerald ring', Gollancz, p. 93, Müller, p.7; Gibson, *WIST* XIV, p. 16; *Liber Prestigiorum Thebidis*, *WIST* XIV, Catalogue 21, p. 173.

CHAPTER 6

Falconry

Adelard's treatise *De cura accipitrum* ('On the Care of Falcons') appears to have been written at the same time as *Quaestiones*, during an interval in more serious considerations.[1] It is the earliest western European treatise on falconry of its kind and is unique by virtue of its information about Anglo-Saxon practices, particularly those described in King Harold's books, as Adelard himself states.[2] These could have been found in a royal library which the Normans acquired at the time of the Conquest. It is another indication of an association by Adelard with the court. It should be remembered that William the Conqueror retained in his service in Somerset the falconer who had previously served King Edward.[3]

The reference to King Harold's books is the first suggestion of a royal Norman library. It is interesting that the Anglo-Saxon sources were books (*libri*). The records of the Exchequer which began in Henry I's reign took the form of rolls. It has even been suggested that Adelard and the abacist called Turchil might have influenced the choice: Adelard because of his experience in Syria, Turchil because he dedicated his text to Simon de Rotol (Simon of the Rolls?). This suggestion is fairly tenuous. Simon's name could just as easily have meant 'of Rutland'. Whether the rolls were started in 1100–10 or 1120–30 is not known. It is a great pity that earlier records than the Pipe Roll of 1130 have not survived, so there is no positive way in which to establish any exact dates earlier than this for Adelard's connection with the court.[4] To be knowledgeable about falcons, however, was a useful asset in a royal household, and access to the books may have provided Adelard with the opportunity he needed to make a noteworthy contribution other than to the Exchequer soon after his return from abroad. Or, indeed, if as some believe, Adelard was a wealthy young man who was tutoring a younger member of his own family, the information on falconry would be a popular change from more serious subjects.[5]

Two texts of Adelard's treatise have survived. In the Clare College MS at Cambridge it is part of a compilation prepared for the son of Emperor Frederick II, a great falconing enthusiast in the thirteenth century. In the Vienna (Wien) MS the work is called 'Hic est de avibus tractatus'. As with most early treatises on falconry there is far more information about the ailments to which hawks are subject than about hawking itself. The format is again dialogue between Adelard and his nephew. This time Adelard begins by suggesting that the intellect is like a bow and grows slack if not kept under tension. As a change from discussing natural causes he will tell his nephew what he knows about the care of falcons and about English practice. His nephew must ask the questions. Although they are relaxing, they will keep their minds active.[6]

The nephew's first question, the point at which the Clare manuscript begins, is 'What qualities are essential to the falconer?'. Adelard replies that the falconer must be sober, patient, chaste, of sweet breath and free of attachments. Sobriety is important because drunkenness is the mother of oblivion, patience because anger will lead to injuries, chastity because association with prostitutes may infect the falconer and thus his charges with parasites or the scab. Bad breath can cause infection and might lead to the birds getting jaundice, or rheumatiz (another name for frounce, a disease which affects the beak of a hawk). The falconer must be free to devote his entire attention to his birds and be ready to take action at any time on their behalf. He should, therefore, not have other responsibilities.[7]

There are brief comments in both texts about taking young birds from the nest and the sort of enclosure or cage to be used for them until they are fully fledged. More detail is given in the Vienna manuscript. Young birds should be taken when they are seven days old, early in the morning when parent birds are absent and before the young themselves leave the nest. At seven days the birds' faculties have developed but they have not yet begun to learn from their parents. The feathers have developed sufficiently for it to be safe to handle them. The *firma*, or cage, must be in a quiet and shady place. The young falcons are fed on flying insects, sparrows, chicken, sheeps' hearts, eggs – but not too often until they are fully feathered. Sinews and tendons are forbidden. The meat is cut into little pieces and spread out above the perch so the birds can eat often. Hunger will debilitate the feathers and it is important that they develop properly. Eggs must be hard-boiled and shells removed, and given mixed with milk. If a bird dislikes something, it should be given at night mixed with a little blood. When the birds have lost their early down they can be exposed to a little more sun. The falcons' house should always be half in the shade. There must be water for the birds to bathe in. They cannot be taken from the cage until their feathers are fully established.[8]

The nephew asks how one knows when the right moment has arrived. Adelard states that it will be after eight days. Each bird is then dealt with individually. Training is about to begin. (Since it is customary to refer to a falcon as 'she' regardless of real sex, I shall follow the practice.) Adelard's subsequent advice describes how she is first removed from the cage. She has to be handled, fitted with jesses to which a lead can be attached (jesses are the leather straps attached to the legs by which the birds are tethered using a swivel and leash at a later stage). She will no longer be kept cooped up but will be tethered to a perch indoors or out. She has to learn to take food only from the falconer. Stage one is to starve her for three days and then to take her from her perch in the dark. To this end the falconer must observe where she is sitting while it is still light.[9]

There are different methods of approach. Some use a cloth over the bird's head and then take her by the legs. Adelard gives the Anglo-Saxon technique whereby the falconer

1 The Liberal Arts – subjects upon which medieval academic studies were based – portrayed in sculpture. (*right*) Dialectic and Aristotle (with Gemini and Pisces alongside); (*below left*) Geometry, Euclid and Rhetoric; (*below right*) Arithmetic, Boethius, Astronomy and Ptolemy. Twelfth century, portail royal, Chartres Cathedral. Courtesy La Crypte.

2 *Astronomia*, one of the Liberal Arts, portrayed on a twelfth-century champlevé enamel casket. She is seated with her back turned, holding up an astrolabe. Courtesy the Trustees of the Victoria and Albert Museum.

3 Seal of Bath Abbey, 1159–1175. The matrix probably dates from the refounding of the Abbey in the tenth century. British Library, Harl. ch. 75 A 30.

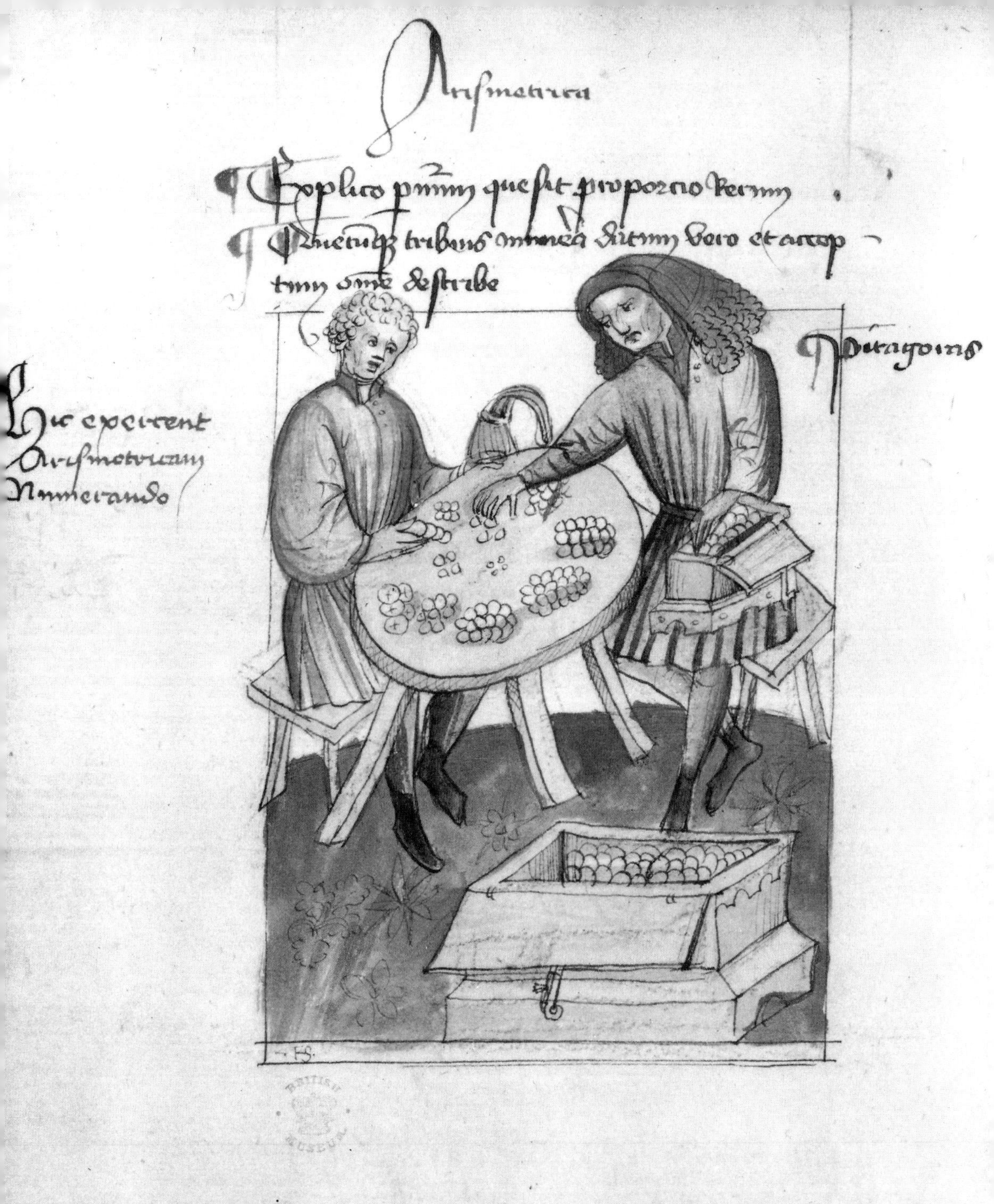

4 *Arithmetica.* The counting method illustrated in this fifteenth-century manuscript is similar to that used in Adelard's time. Courtesy British Library, BL Add. MS 15692, f.28v.

5 Falconer in pursuit of cranes. In *De cura accipitrum* Adelard recommends that falcons be bathed in the water in which a crane has been cooked. This initial is from a twelfth-century manuscript from Cîteaux, 'S. Gregorii Magni, Moralium in Job', MS 173, f.174. Courtesy Bibliothèque Municipale de Dijon.

6 Bridge at Misis, modern Turkey, partially destroyed in 1114 as the result of an earthquake. Adelard mentions this bridge and the earthquake in *Quaestiones naturales.* According to Dr John H. Harvey, the bridge was repaired in the twelfth century using pointed arches alongside rounded ones. Courtesy Dr John H. Harvey.

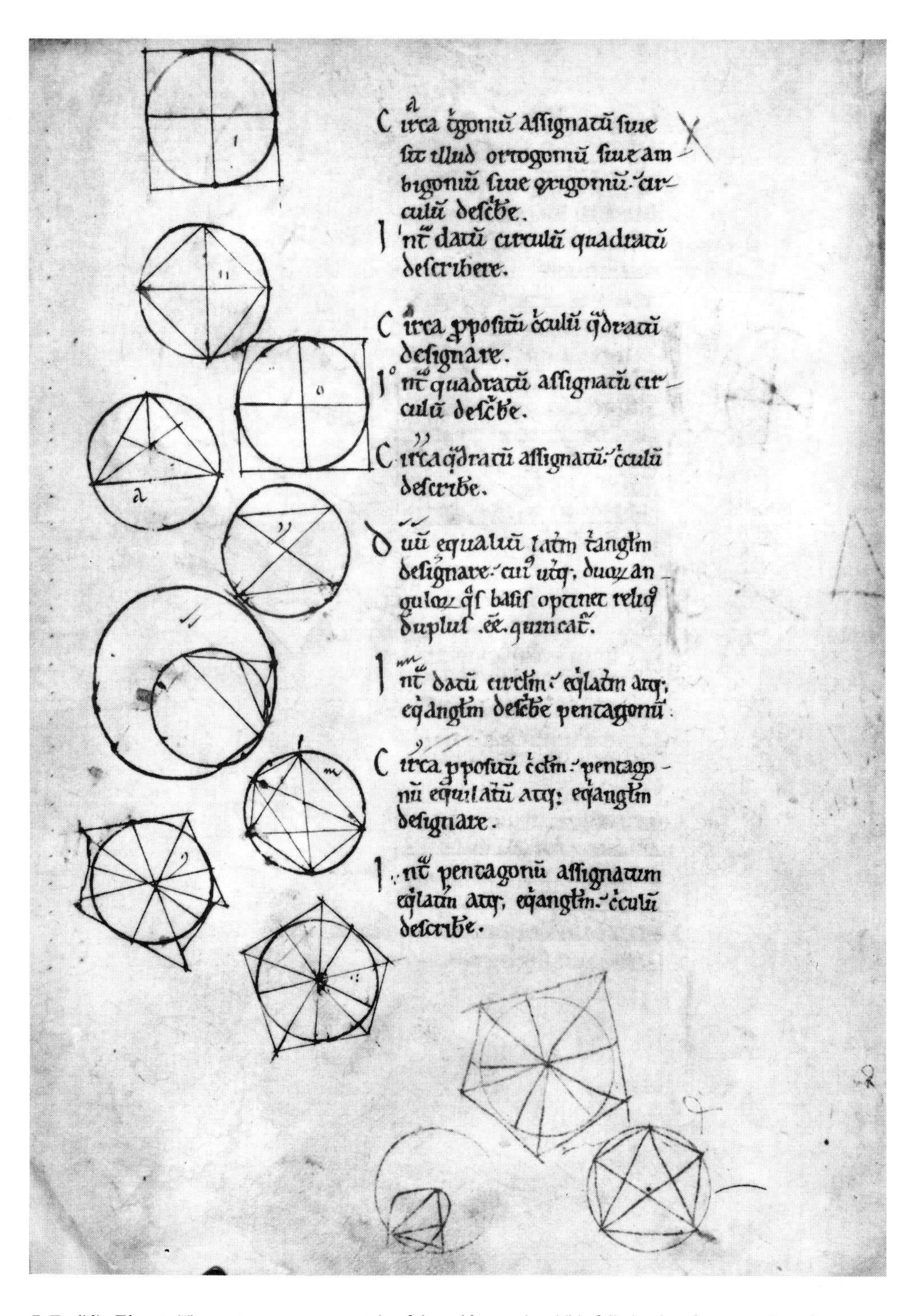

7 Euclid's *Elements*. The pentagon was one matrix of the golden section. This folio is taken from a version of Adelard's translation of Euclid which is also associated with Ocreatus, one of his pupils. Courtesy British Library, BL Royal 15A XXVII, f.11v.

8 European astrolabe, *c.* AD 1195, front. The front of an astrolabe incorporates inside its rim the 'Womb' or the 'Mater' into which a climate plate appropriate to the latitude of the observer can be inserted. Above this, the rete with its star pointers is designed so that it can be revolved clockwise once in the sidereal day. The photograph shows the position of the stars at noon when the sun is in the first degree of Capricorn. The North Pole is at the centre of the instrument; the meridian due south is at the top. The star pointers near the top are Altair, east of the meridian; Ras Alhague (Caput Serpentarii), west of the meridian; Vega, near the zenith. The label, which is used to line up the star pointers and to indicate the time, is missing on this astrolabe. Courtesy the Trustees of the British Museum, MLA 1961, 12-1.1.

9 European astrolabe, *c.* AD 1195, back. The rim shows the 360° of the circle, the signs of the zodiac, with thirty degrees for each sign and the months and days of the year correlated with the degrees of each zodiac sign. To set the astrolabe correctly for any one day it is necessary first to select the correct degree of the zodiac in relation to the calendar and then to turn to the front of the astrolabe and rotate the rete until that degree corresponds with the time of day. Alternatively, using the back of the astrolabe the sighting device called the alidade can be used to find the direction of a star and thus enable one to adjust the rete and find the local time. The shadow squares on the back are used for surveying. Courtesy the Trustees of the British Museum, MLA 1961, 12-1.1.

10 Hispano-Moorish astrolabe, AD 1026, front. The rete here is a thirteenth-century replacement and is set similarly to the astrolabe in Plate 8. Both photographs demonstrate how precisely the star pointers can establish the co-ordinates of a celestial position on the climate plate underneath. Courtesy the National Museums of Scotland, T1959.62.

wears his special gloves and grasps the bird across her back from above. In this way the danger of damage to the wings from flapping is reduced. According to Swaen, the editor of the treatise, this is the most interesting comment on falconry itself, clearly indicating a different method used by Anglo-Saxons in the management of their birds.[10] Having fitted the necessary equipment the falconer proceeds with the long process of taming and training, persuading the falcon to sit on the left wrist while stroking her with a rod or a long feather, about 15 cm (6 in) long, held in the right hand.

It seems that it was Anglo-Saxon practice to carry the falcon on the left wrist and it is sometimes said that this practice differed from that on the Continent. Among the many occasions when falconers appear in art, as in the Bayeux Tapestry, there seems to be no general rule as to which wrist was favoured. Harold is shown after his capture, his hawk on his left wrist, but so also is Guy de Ponthieu who is leading him.[11]

Care must be taken that the bird does not fly off until the falconer knows she will return. This is done by providing the bird's food at a regular time and always in the same place. As soon as she 'plumes' or plucks her first whole sparrow, she is ready for the next phase of her training, which is called 'making in'. Like all stages in the lengthy process, this must be accomplished without haste – *leviter accede* in Latin.[12]

The information which Adelard gives varies a little from the practice outlined in standard writings of a later date, but not significantly. It is, however, very compressed, probably because all who were interested at the time knew the sort of detail about which we require explanation. It is difficult for us to visualise exactly what Adelard is talking about. One of the essential aspects of taking eyases from the nest is that exactly the right moment must be chosen. Too soon and the birds are difficult to rear, prone to become screamers, subject to cramp, which can prove fatal. If the young birds have left the nest, it is harder to capture them without injuring the feathers. Once taken, young falcons are usually kept in a roughly made nest on the floor of a large loft, shed or coach house. They are fitted with jesses and bells. Different colours are used for jesses so that the birds can be identified . Young birds are left to themselves, with the falconer showing himself as little as possible to avoid their associating him with their food – which will again induce a screaming habit.

According to Adelard's method the falcon is trained to the wrist as soon as she is fully feathered. The practice of starving the bird to make her eager for food is no longer deemed necessary. Some falconers did not train birds to the wrist until after the period called the 'hack' when a bird is taken into the open, trained to come to a board or a lure for food and then allowed to fly off freely to develop her natural flight. Food is always tied down to a board so the bird has to eat *in situ*. A lure is weighted so that the bird finds it too heavy to take away (a feather concealing meat and weighted by something like a horseshoe is an example).

As soon as young hawks have spent a day on the wing and returned to feed at evening it may be considered safe to allow them a state of liberty until they learn to find prey for themselves; this will not be for some weeks. Food must be provided regularly at the same time and place, and a decision made at the outset as to whether the bird is to be fed from a board, from a lure, or from the falconer's wrist. Directly a hawk absents herself at feeding time it is a sign that she is killing for herself, and if missing a second time must be caught. A sort of bow net is used for this.

Serious training now begins: the bird is hooded and the swivel and leash attached to the jesses. She is carried for several hours at a time on the wrist and stroked with a feather until she is accustomed to it. The hood is so designed that she can eat while wearing it but during training she is gradually accustomed to eating without a hood.

As Adelard tells us, the bird is first handled in the dark, so she is first fed without her hood by candlelight. When this causes no problems she can be tried in daylight. It is necessary always to continue feeding her after the hood is replaced until she is used to it; otherwise she will associate the hood with the absence of food.

Hooding a bird well is a tricky operation. The falconer needs to become deft at placing the hood on the bird's head with one hand and tipping her forward slightly with the other so that it slips on easily.

Adelard does not describe the final stages of training. The lure is thrown first for short distances so the bird flies to it along the ground. It is then thrown into the air. The intention is gradually to accustom the falcon to seizing prey above the falconer's head. Once killed the prey must be released to the falconer who feeds the falcon as her reward. Training the falcon to what is required is a long and patient process. 'Making in' is successfully accomplished when the bird uses natural instinct to kill but does so in circumstances which suit the falconer and to satisfy him rather than her own appetite.[13] In response to questions from the nephew Adelard now turns to various ailments and their treatment so no further details on actual hawking are presented. The list of afflictions begins with those involving the head – and first to be dealt with is frounce, also called rheumatiz and fungus. A present-day falconer describes it as follows:

> For a long time one of the major killers was 'frounce', which unless caught in the very early stages is nearly always fatal – the onset can often be traced to pigeons and is due to a protozoan parasite *Trichomonas gallinae*, first described by Rivolta in 1878 – the disease is first apparent in the form of a cheesey like growth on the tongue.[14]

For this complaint Adelard suggests that little pills be made up of salt and pepper in butter and inserted in the bird's nostrils, so clearly one effect of the disease is that the nose is congested. He recommends that warm poultices be applied to the head or the head be dipped in warm water up to the eyes or ears. Warm stones near the feet are proposed, presumably to keep the bird warm.

Again the Vienna manuscript gives more detail, this time about the difference between dry and moist frounce. For moist you use mead and honey mixed in a little bowl, and a linen cloth is then soaked in the mixture and placed in the bird's mouth which is then shut. It is sometimes necessary to cut away the growth and then apply the medicament.[15]

If the falcon vomits she is offered water and then later fed mice, or sparrow, or meat which has been boiled in water.[16]

For accidents to the chest or when the falcon's breathing is affected the suggested cure is iron filings(!) mixed with dove's heart or radish and butter. The butter is likely to be May-butter, sweet butter kept for medicinal reasons.[17]

A disease called *inflatio* is treated with 'aquilegia' (Latin *aviligia).* Although 'aquilegia' seems the obvious translation it is likely the herbal remedy was made with meadow-sweet. Lemon balm, for which this Latin word was also used in Adelard's time, was not a native plant.[18]

Adelard now comes to the problem of lice – where the feathers must be doused with wormwood juice and it is suggested that a woollen thread be coated with goose or chicken fat and then tied loosely around the neck and legs. It is further recommended that a new perch be made from spindlewood (*fusenum*) and that it be moved to a different position. Absinth or wormwood juice is by tradition known as a disinfectant so this seems reasonable. A somewhat similar treatment is recommended for mites, using a mixture of soda and ash bark or wormwood vinegar and salt boiled together to treat the place where mites are found.[19]

At this point in the Vienna manuscript there is a blank so it is not certain for what disease the use of red dock is recommended. A modern herbalist gives several suggestions, for example that externally it is used for eruptive and scorbutic diseases.[20]

Jaundice is next to be discussed. Surprisingly this is another serious hazard for falcons, and Adelard describes it as incurable if it causes 'rupture' – though this may be a mistranscription for an indication of the advanced states of the disease. The plants recommended for treatment are either ivy (*Hedera helix*) or brownwort (water figwort), or *Helleborus niger* in the Vienna manuscript. St John's wort, stonecrop and leek are offered as possible cures in the Clare manuscript. St John's wort is also known as a source for the treatment of kidney disease in humans. In the case of falcons the bird is offered meat prepared with the appropriate herb.[21]

A falcon is feverish when her eyes close as she sits on her perch and her head appears swollen. The remedy is a herb recorded as '*lenticulum fontis*' which is ground to a powder and sprinkled over the bird's food. *Lenticulum fontis* is not lentil but *Lemna minor* (duckweed). A German scholar calls it *Wasserlinse.* A further proposal is that falcons should be bathed in water in which a crane has been cooked. The word for crane, *grus*, appears in both manuscripts. Apparently cranes were popular quarry for medieval falconers and nested in Britain until the end of the sixteenth century. It is exciting that after 400 years they have started to do so again.[22]

A mite or insect which attacks the falcon's feathers and causes feathers to fall out is then discussed. Chicken meat treated with savory or creeping cinquefoil given in the gut of a dove is prescribed for this. For ticks and lice dip meat in a bowl in which goat's milk has been boiled and then combine it with broom (*genista*) and incorporate it in the gut of a hen. Medicines were offered in the intestines of hens or doves. It seems odd that an internal medicine is suggested for mites and ticks.[23]

For 'the stone' the juice of *lithospermum* (gromwell) is the remedy. For constipation (obstructed bowel) meat should be dipped in warm water – this remedy continued certainly into Victorian times. One of the falconer's priorities was keeping a watchful eye on the falcon's castings, the pellets excreted by the bird which contained the bits of fur or feathers deliberately fed as the falcon's natural diet. If these showed unusual symptoms the use of one of the many remedies was indicated.[24]

The disease called *bisticus* seems to be a major blocking of the bowel and a concoction of mallow and savory with pork fat given warm will cause the anus to foment and resolve the problem. The Vienna manuscript offers other suggestions – a mixture of *Inula helenium*, known medicinally for treating chest ailments and a possible cure for tuberculosis is one. So there may be a confusion here with phthisis or *tesga* (consumption) which also attacks falcons. *Tesga* as a disease is mentioned again in a final section of the Clare manuscript which is not in dialogue form and may perhaps not have been written by Adelard. This time the word is used for a disease of the bowel and then shortly afterwards it is used again to describe a severe cold in the nostrils and again in connection with a treatment suitable for ticks. The problem of interpreting these manuscripts where the scribes knew little or nothing about falcons is a difficult one as the editor has stressed.[25]

After *bisticus* Adelard's treatise considers complaints to do with the falcon's feet and legs – warts and gout. For warts the roots of blackthorn or plum trees are soaked in vinegar and tied around the growths so they can be drawn out when they have softened. For gout the barks of ash, apple, oak and blackthorn are boiled, allowed to cool and applied with soap.[26]

Adelard's nephew asks now about special problems during the period when the falcon moults and may be off colour. Adelard replies that the falcon should have three hours a day in the sun and a varied diet. Mice are a suitable food, so are young cranes and ground-up animal intestines. Meat should be sprinkled with powdered honeycomb. A snake should be cooked and the broth fed to chickens which are then used to feed the falcons. Adelard gives a dissertation on the importance of diet and the distinction to be made between fat and lean meats and their properties in ensuring that if the falcon becomes too fat during the moult the right food is given to restore her to health.[27]

One of the interesting features of the treatise is the practical medical knowledge which it contains. The section on how to feed a falcon at the end of a moult is an example. The bird has had little exercise and if she is then allowed a very sudden burst of energy it may cause heart failure. Adelard's explanation indicates that it is wise to avoid fatty meat in the bird's diet during the moult if the bird begins to get fat.[28]

What to do to persuade a reluctant falcon to fly is considered also. This involves keeping the bird hungry and then tempting her with a particularly appetising morsel of pork tongue.[29]

The final section of the Vienna manuscript differs from the final part of Clare and gives a long comment from Adelard on Anglo-Saxon recipes for a mild tonic and a more severe purgative. The section begins with the word *Salva*, probably the equivalent of the 'Save' in many herbals. The mild remedy is made with betony, wild thyme, mallow and woodruff. Plaintain and yarrow are cooked in butter, in equal quantities, and then the other ingredients are added; what is not used is reserved for another occasion. The more severe purgative contains parsley, betony, thyme, again cooked in butter. The root of inula is added. It is mixed with mead and given without any meat.[30]

Adelard completes the treatise by saying that he has recorded what he knows of falconry but will not be offended if other information is added. His work contains a detailed and quite specific summary of problems and remedies as well as his abbreviated account of the taking of eyases from the nest and early training. There is so little about the actual sport that it must be assumed that this was very familiar to those for whom Adelard was writing. It could possibly have been for Henry I himself. He was interested in animals and even had a royal menagerie at Woodstock. At the time Adelard recorded all this information he may well have been seeking further patronage.

It is also very obvious that Adelard had a keen interest in plants and their uses. For us today looking back 900 years it is surprising that all the plants mentioned except *satureia* (savory) still grow wild in the United Kingdom.[31]

Notes

1 Adelard of Bath, *De cura accipitrum*, ed. A.E.H. Swaen (Amsterdam, 1937), Introduction, p. iii. Contains Latin texts and detailed notes on two manuscripts: Vienna (W) and Clare College, Cambridge (C), *WIST* XIV, Catalogue 4, p. 167.

2 *De cura accipitrum*, W, p. 1, l.12. See also C.H. Haskins, 'King Harold's Books', *EHR* 37 (1922), pp. 398–400; Judith A. Green, *The Government of England under Henry I* (Cambridge, 1986), p. 161.

3 See above, ch. 1, p. 3; Robin and Virginia Oggins, 'Some Hawkers of Somerset', *PSANHS* 124 (1980), pp. 51–3.

4 *De cura accipitrum*, W, p. 7. ll.11–13, *Haroaldi [sic] regis libri*; M.T. Clanchy, *From Memory to Written Record, 1066–1307* (London, 1979), pp. 108, 110. See also ch. 1, n. 34. As to the origin of Simon de Rotol, Harvey (letter to L. Cochrane, 19 October 1990) points out that Rutland, the later county, was a soke of the Queens of England, and so linked to the court. Ratley, Warwickshire, is also notionally a possibility.

5 Gibson, *WIST* XIV, p. 10.

6 *De cura accipitrum*, W, p. 7, ll.4–15.

7 Qualities of falconer: *De cura accipitrum*, C, p. 1, ll.2–7; W, p. 7, ll.4–15.

8 Taking of birds from nest, early feeding: *De cura accipitrum*, C, p. 1, ll.8–25; W, pp. 7–9, ll.27–66.

9 Taking birds from perch, early training: *De cura accipitrum*, C, pp. 1,2, ll.26–36; W, p. 9, ll.66–78.

10 Adelard's description of Anglo-Saxon method: *De cura accipitrum*, W, p.9, ll.82–7, notes, p. 24; C, p. 2, ll.36–40; W, p.9, ll.78–87.

11 Christian Antoine de Chamerlat, *Falconry and Art*, with a foreword by T.A.M. Jack, Sotheby's Publications (London, 1987). Reproduction of the Bayeux Tapestry, p. 85. The illustrations in this book indicate that it was equally usual for falconers on the Continent to support their hawks on the left wrist, although in a number of cases the right fist is used instead.

12 Taking care lest bird fly off: *De cura accipitrum*, C, p.2. ll.40–3; W, p. 9, ll.87–100. 'Making in'.

13 E.B. Mitchell, *The Art and Practice of Hawking* (London, 1900), pp. 27, 59–62, 95, 96; Gerald Lascelles, 'Falconry', in Harding Cox and Gerald Lascelles, *Coursing and Falconry* (London, 1892), pp. 236–46.

14 Allan Oswald, *History and Practice of Falconry* (Jersey, 1982), pp. 107, 108.

15 Frounce: *De cura accipitrum*, C, p.2, ll.44–54; W, pp. 9, 10, ll.100–51.

16 Vomiting: *De cura accipitrum*, C, p. 2, ll.54, 55; W, p. 11, ll.151–6.

17 Chest ailments: *De cura accipitrum*, C, p. 2, ll.55, 56; W, p. 11, ll.155–63.
The use of iron filings is confirmed by Gunnar Tilander, *Grisofus Medicus, Alexander Medicus* (Lund, 1964), p. 14, para. 5. For radish (*raphanus*) see George Usher, *A Dictionary of Plants Used by Man* (London, 1974), p. 497; W. Keble-Martin, *The Concise British Flora in Colour* (London, 1965), pl. 11. Harvey suggests this to be a variety of *Raphanus sativus* (letter to L. Cochrane, 7 March 1990). *De cura accipitrum* (C, p. 17, n. l.57), Swaen also gives horse-radish in translation but this was not an English plant.

18 *Inflatio*: *De cura accipitrum*, C, p. 2, ll.58–61; W, p. 11, ll.159–63. The complaint is probably flatulence. Swaen suggests (p. 17, C, n. l.59) that the plant in question is columbine, but Harvey points out that in the early twelfth century 'it is almost certain that it meant

Meadow-sweet (*Filipendula ulmaria* (L.) Maxim.); and virtually certain that it did *not* mean columbine. There is, however, also confusion with Lemon Balm (*Melissa officinalis* L.). Meadow-sweet and Balm were both diuretic so the medicinal use presents no difficulty' (Harvey, 7 March 1990). Balm was not a native plant so meadow-sweet is the likely herb. For properties of these see Usher, p. 257; Keble-Martin, pl. 26. Properties of columbine (aquilegia), Usher, p. 53; Keble-Martin, pl. 4.

19 Lice, mites: *De cura accipitrum*, C, p.2, ll.58–66; W, p. 11, ll.163–74; wormwood (*Artemisia absinthium*), Usher, p. 61; Keble-Martin, pl. 46. Spindle tree (*Euonymus europaeus*), Usher, p. 551; Keble-Martin, pl. 20.

20 Red dock (*rumex*), Usher, p. 514; Keble-Martin, pl. 74. For information on medicinal uses of red dock or water dock see Mrs M. Grieve, *A Modern Herbal* (London, 1931), pp. 259, 260. 'Red dock was recommended by Platearius for the treatment of mange; by Macer because it "comforts the stomach and destroyeth wind"' (Harvey, 18 October, 1990).

21 Jaundice (also rupture, hernia?): *De cura accipitrum*, C, p. 3, ll.69–75; W, p. 12, ll.178–93; ivy (*Hedera helix*), Usher, p. 297; Keble-Martin, pl. 41. The Latin text here refers to a plant which the Anglo-Saxons call *nigram herbam*. Swaen suggests (n., p. 26) that this is the Anglo-Saxon name for betony, modern brownwort, but the plant which would best suit the description is *Scrophularia nodosa* or *Scrophularia aquatica*, unless *Helleborus niger* is meant. Usher, p. 531, for *Scrophularia nodosa*; Grieve, p. 314, for water figwort, also called brownwort and water betony; Keble-Martin, pl. 63. For brownwort see also *Shorter Oxford English Dictionary*, 3rd edn, vol 1, p. 227. *Helleborus niger*, Grieve, p. 388; Usher, p. 299; Keble-Martin, pl. 4 (shows *Helleborus foetidus* and *Helleborus viridis* which are used as substitutes). See also Nicholas Culpeper, *The Complete Herbal to which are added the English Physician and Key to Physic* (repr. privately for ICI, 1953), pp. 131, 132; St John's wort, (*Hypericum*), Usher, p. 313; Keble-Martin, pl. 17; stonecrop (*Sedum*), Usher, p. 533; Keble-Martin, pl. 33; houseleek (*Sempervivum*), Usher, p. 534; Grieve, p. 422; Keble-Martin, pl. 33. Literally, *nigra herba* should mean 'black grass', i.e. black medick (*Medicago lupulina*). Harvey has not found it as = betony or water betony. *Helleborus viridis* was the British native species regarded as 'black hellebore' in the Middle Ages (Harvey, October 1990).

22 Fever: *De cura accipitrum*, C, p. 3, ll.75–9, p. 18, n. l.76; W, p. 12. ll.193–8; duckweed (*Lemna minor*), Usher, p. 348; Keble-Martin, pl. 88. Translation confirmed by R.E. Latham, *Revised Medieval Latin Word-list* (London, 1965), p. 263.
Michael McCarthy, Environment Correspondent of *The Times*, reports, 24 April 1990, that the crane has returned to Britain to breed after 400 years.

23 Mites and ticks: *De cura accipitrum*, C, p. 3, ll.79–85; W, pp. 12, 13, ll.198–209; savory, (*Satureia*), Usher, p. 527 (not illus. in Keble-Martin); creeping cinquefoil (*Potentilla reptans*), Usher, p. 480; Keble-Martin, pl. 27; broom (*Genista*), Usher, p. 271; Keble-Martin, pl. 42.

24 Stone: *De cura accipitrum*, C, p. 3, ll.85, 86; W, p. 13, ll.209–13; gromwell (*Lithospermum*), Usher, p. 359; Keble-Martin, pl. 61. Gromwell used for treating 'gravel': *Shorter Oxford*, vol. i. p. 834. See also Francis Henry Salvin and William Brodrick, *Falconry in the British Isles* (London, 1855, repr. 1980), pp. 126, 127.

25 *Bisticus*, *De cura accipitrum*, C, p. 3, ll.88–92; W, p. 13, ll.215–24; mallow (*Malva*), see Culpeper, 156–9; Usher, p. 375; Keble-Martin, pl. 18; elecampane (*Inula helenium*), Usher, p. 319; Keble-Martin, pl. 45; Culpeper, pp. 98, 99; *Shorter Oxford*, vol. i, pp. 1037, 1038. For *tesga* see Tilander (n. 17 above), p. 253. Difficulty with plant names as transcribed in texts, *De cura accipitrum*, Swaen's Introduction, pp. vi, vii.

26 Warts, gout: *De cura accipitrum*, C, p. 3. ll.92–6; W, p. 13, ll.224–30; trouble with feet, Oswald, pp. 109, 110; Salvin and Brodrick, p. 128; blackthorn (*Prunus spinosa*, fruits are sloes), Usher, p. 485; Keble-Martin, pl. 26.

27 Moult: *De cura accipitrum*, C, p. 4, ll.101–8; W, pp. 13,14, ll.234–47.

28 Fatty meat: *De cura accipitrum*, C, p. 4. ll.109–14; W, p. 14, ll.247–55.

29 Refusal to fly: *De cura accipitrum*, C, p. 4, ll.114–17; W, p. 14, ll.256–60 (see above, n. 24, *re* castings).

30 Anglo-Saxon remedies; *De cura accipitrum*, *Salva*, W only, p. 14, ll.260–76; detailed notes, Swaen, pp. 27,28. Swaen considers this text to be 'a bit of native falconry'. For a mild purgative: plantain (*Plantago*), Usher, p. 468; Keble-Martin, pl. 71; yarrow (*Achillea*), Usher, pp. 16,17; Keble-Martin, pl. 45; *Shorter Oxford*, vol. ii, p. 2464; betony, Usher, p. 552 (n. 21 above); wild thyme (*Thymus serpyllum*, *Thymus vulgaris*), basil thyme, wild basil (*Calamintha acinos*, *Calamintha clinopodium*), Usher, p. 576; Keble-Martin, pl. 68; *Shorter Oxford*, vol. ii, p. 2187; mallow (*Malva*), (n. 25, above); hawthorn haws, common hawthorn (*Cratageus monogyna*), Usher, p. 182; Keble-Martin, pl. 42; woodruff (*Asperula odorata*), Usher, p. 265; Keble-Martin, pl. 18. For a stronger purgative: parsley (*Petroselinum crispum*), Usher p. 450; Keble-Martin, pl. 37. Betony, wild thyme, inula root, mallow (see above).

31 The following list indicates the plates in Keble-Martin where illustrations of all of the plants except savory (*Satureia*) can be found:

- Pl. 4 Columbine (*Aquilegia*), hellebore (*Helleborus*)
- Pl. 11 Radish (*Raphanus*)
- Pl. 17 St John's wort (*Hypericum*)
- Pl. 18 Mallow (*Malva*)
- Pl. 20 Spindle tree (*Euonymus europaeus*)
- Pl. 26 Blackthorn (*Prunus spinosa*), meadow-sweet (*Filipendula ulmaria*)
- Pl. 27 Cinquefoil (*Potentilla*)
- Pl. 31 Hawthorn (*Cratageus monogyna*), apple (*Pyrrus sylvestris*)
- Pl. 33 Stonecrop (*Sedum*), houseleek (*Sempervivum*)
- Pl. 37 Parsley (*Petroselinum crispum*)
- Pl. 41 Ivy (*Hedera*)
- Pl. 42 Broom (*Genista*), woodruff (*Asperula / Galium odorata / odoratum*)
- Pl. 45 Elecampane (*Inula helenium*), yarrow (*Achillea millefolium*)
- Pl. 46 Wormwood (*Artemisia absinthium*)
- Pl. 58 Ash (*Fraxinus excelsior*)
- Pl. 61 Gromwell (*Lithospermum*)
- Pl. 63 Water figwort, water betony (*Scrophularia aquatica*)
- Pl. 68 Wild thyme (*Thymus serpyllum*), (*Thymus vulgaris*); basil thyme (*Calamintha acinos*), wild basil (*Calamintha clinopodium*)
- Pl. 69 Betony (*Betonica officinalis*)
- Pl. 74 Red dock (*Rumex*), plantain (*Plantago*)
- Pl. 77 Oak (*Quercus*)
- Pl. 84 Leek (*Allium ampeloprasum*)
- Pl. 88 Duckweed (*Lemna*)

See John Harvey, *Mediaeval Gardens* (London, 1981, rev. edn, 1990), list of plants pp. 168–80.

CHAPTER 7

Adelard's Translation of Euclid's *Elements*

During his travels Adelard obtained an Arabic version of *The Thirteen Elements of Euclid's Geometry*.[1] He refers to having translated fifteen books of Euclid in his later treatise on the astrolabe, but two of these were accretions.[2] Euclid's *Elements* was first produced in 300 BC in Alexandria and constitutes one of the greatest intellectual achievements of all time. That no complete Latin translation survived the Dark Ages meant that European mathematics was singularly impoverished. In Adelard's youth only the arithmetic and geometry of Boethius were regularly studied, although Gerbert had extended curiosity about Arab sources. Gerbert's mathematical ideas and analysis were often attributed to Boethius during the eleventh century; Adelard's familiarity with them has been mentioned.[3]

The Arabs, in contrast to the West, had been acutely aware of Euclid's importance and there were two translations from Greek into Arabic, one by al-Hajjaj (al-Hajjaj b.Yusuf b. Matar, fl. 786–833) and another by Ishaq b. Hunayn (d. 910, 911). The second was edited by Thābit b. Qurra (d. 901).[4]

A great deal of scholarly attention has been given to Adelard's translation of the *Elements*. Because of his interest in Euclid's geometry, Adelard has also been referred to occasionally as translator of Euclid's *Optics* and *Catoptrica*.[5] Usually any reference to Adelard's Euclid is a reference only to the *Elements*. Three versions were differentiated by Marshall Clagett: in his view Adelard I is a translation probably from an Ishaq-Thābit text. The argument in favour of Adelard as translator is the inscription in one manuscript – there are five altogether.[6] Adelard II became the popular text – there are fifty-six manuscripts. There is general agreement that this version was definitely written by Adelard. It has been shown that he used an Arabic text in some instances and then replaced it with a text from the Boethian tradition. Adelard II was used by Campanus and became the basis for the first printed text.[7]

There is a possibility that a student of Adelard's, John or Nicholas Ocreat, assisted in the preparation of the texts as there is a reference to him in some of the manuscripts. Authorship of one (British Library, Royal 15 A. XXVII), was attributed to Ocreat in Bernard, a seventeenth-century catalogue at the British Museum, but the folio has now been lost.[8] Ocreat also dedicated a treatise on Arabic arithmetic to Adelard, describing him as his master. It is thought likely that Ocreat grafted some of his own ideas onto a version of Adelard II. Adelard II contains many Arabic terms and shows that an Arabic text was at hand when it was written. The text also indicates that Adelard was influenced by the more common pseudo-Boethian propositions which still circulated.

Adelard III enunciates propositions as in Adelard II but proofs are given in a complete, specific and formal way, not unlike the translation of Adelard I. Adelard III is the version which was used by Roger Bacon in the next century. Although the translation was Adelard's, additional material could have been introduced perhaps by Jordanus of Nemore. Because pseudo-Boethian geometry had enunciated propositions without demonstration there was a prevalent idea that Euclid had been the author only of definitions, postulates and enunciations, and so demonstrations were considered commentary. This notion persisted into the Renaissance: the title *Euclidus liber Elementorum ... translatus ab arabico in Latinum per Adelardum Bathoniensem sub commento Magistri Campani Novarriensis* indicates this. The demonstrations which were part of the original text were attributed to Campanus.[9]

Shortly after Adelard completed his translation Hermann of Carinthia and Gerard of Cremona also made translations from the Arabic. These did not have the success that Adelard's had. Detailed studies of Arabic-Latin translations continue and careful comparison of the texts has revealed a great deal of information.[10]

Adelard's translation of the *Elements* was of profound importance for the development of scientific thinking generally since it made Europeans conscious of the method by which theorems were proved deductively. Increasing interest in mathematics ensured that it became widely known. Adelard's translation became the basis of all editions in Europe until 1533.[11]

The appearance of Adelard's Euclid coincides with the beginnings of medieval Gothic architecture. There is good reason to believe that one strengthened the other. A number of scholars support this view, in particular John Harvey, the architectural historian, and Christopher Brooke, an authority on the twelfth-century renaissance.[12] They do not, of course, claim that Gothic style owes anything specifically to Adelard but believe that the greater accuracy in the setting out of foundations and the construction of buildings of greater span and loftier height must have been based on a greater knowledge and use of geometry. This accelerated the transition from the rounded Romanesque arch to the pointed Gothic one.

Two further reasons are cited by Ivor Bulmer-Thomas for believing Adelard's influence to have been of prime importance. One is the result of a careful study of pre-Norman churches in England, so many of which have skew chancels, a fact which shows that builders found it difficult to achieve true rectangles. Added to this is the fact that rounded doorheads were often not true segments of a circle. Moreover, during the Middle Ages a strong tradition arose amongst Freemasons that Euclid was the founder of their order. Two West Country manuscripts refer to these early origins.[13]

With Adelard's translation a new differentiation between the practical geometry known previously and the theoretical geometry of Euclid developed. Hugh of St Victor produced a practical geometry about 1125, summarising the principles which were used for building and measurement. The instruments to be employed were the astrolabe, the right triangle, several staffs and rods, a mirror and a gnomon. There is no mention of the quadrant whose function must have been absorbed in the use of the astrolabe.

In Adelard III practical geometry is included in liberal practice, part of the practical side of the division of knowledge. With respect to geometry as one of the liberal arts, Adelard distinguishes between two branches in terms of the functions and the tools of the master. The master is both demonstrator and a practitioner. As demonstrator he must explain the theorems; as practitioner he must make actual measurements. As demonstrator he uses a stick and a table covered with sand, as we might use a blackboard; the instruments of the practitioner are the measures of geometry, namely a pair of compasses. It is not made clear whether Adelard's measurer was working practically or pedagogically.

Practical geometry, according to Hugh of St Victor, involves the use of ratios. Most of the measuring technique which he describes depends on proportion – the distance to be measured as a certain proportion of a known distance: that ratio is the same as one determined by using an instrument. The resulting proportion is used to compute the unknown distance. The ratios most used are those of right triangles sharing the same relation to the hypotenuse, as they are explained in *De eodem* and *Mappae clavicula.*

Euclid's *Elements* is the model textbook of theoretical geometry, and one would not necessarily expect to find much concern with its practical aspects on the part of translators. Adelard, as represented in Adelard III, however, has gone to great pains to relate the whole geometry of commensurate magnitudes to the science of practical measure, thus associating his new discoveries with his former work.[14]

From this point of view Adelard's Euclid could indeed have had an impact not perhaps directly on builders but on the men who first commissioned them and then on the master masons. Once techniques of manipulating geometric forms had been devised, they are said by some to have become the so-called 'secrets of medieval masons' and no great mathematical genius was involved in carrying them out.[15]

It would be wrong to exaggerate the role of the newly available geometry in the development of the Gothic. The outstanding characteristic of the new style was the use of the pointed arch, and it is generally accepted that contact with the Saracens first introduced the pointed arch to the West. Harvey quotes Christopher Wren's comment that the new architecture should be called Saracenic rather than Gothic. It is impossible now to distinguish how much the improvement in building techniques was due to superior building methods learned from the Saracens and how much to greater emphasis on geometry. Certainly the transition was rapid following the First Crusade.[16]

It should be recalled that the building of purely Gothic churches had been preceded in the eleventh century by the occasional use of pointed arches alongside round. This happened at Monte Cassino and the idea was pursued at Cluny. The first impetus towards a new style had come with the defeat of the Moors in Spain and the conquest of Sicily. But it was the use of captive master masons for European buildings which strengthened the development. One of these was Lalys who designed Neath Abbey in Glamorgan in 1129; he is said to have become architect for Henry I. Since the leading feature of the new style was the pointed arch with its visible heavenward thrust, it is not surprising to find it identified with the new Cistercian foundations inspired by the leadership of St Bernard of Clairvaux.

With the completion of Abbot Suger's new church of St Denis outside Paris and the rebuilding of Chartres after a fire in 1145, Gothic architecture began to assume prominence throughout Europe. By this time Adelard's Euclid had become available to students in the cathedral schools.[17]

In tracing sources for the pointed arch it is clear that it was not as prevalent throughout the Arab world as is sometimes assumed. That is why Harvey, who noted the similarity in design between a mosque at Bitlis in Anatolia and the Abbey of Fontenay, considers the coincidence that Adelard may have seen the bridge at Misis being repaired of some importance. A link between Fontenay and the England of Henry I can be established through the patronage of Fontenay by Everard, Bishop of Norwich, who contributed financially to its construction. The coincidence that Henry's Queen, Matilda, commissioned a bridge across the Lea at Stratford has already been mentioned (Chapter 4, p. 36). This structure was unusual enough to give the name of Bow to the place. The bridge was rebuilt in the fifteenth century when it certainly had segmented arches, but whether they were there originally is problematic.[18]

The pointed arch was useful because of its versatility. The width of the arch could be varied by a considerable amount while its height remained constant. Thus it solved the builder's chief difficulties. The wall arches could be narrow and sharply pointed, and the diagonal rib, perhaps half as long again, could be a true semi-arch. All distortion of arches could be avoided, and the use of carefully designed stone webs enabled the reduced weight of the roof to be transmitted onto columns, piers and buttresses while the walls could be pierced for windows.[19]

The arrival in the West of Euclidean geometry was the last stage in a process of development in architecture which lasted for three generations, so must be kept in perspective. The first phase was the attempt to build churches on a really large scale which meant achieving a considerable roof span. The next step was rib vaulting as at Durham, which was also indebted to a technique observed in Spain and the ideas which had come from the East. Gothic and the pointed arch was the next stage – the discovery that it was possible to erect a sort of scaffolding of stone to reinforce slim pillars and narrow ribs.[20] The Cistercians had their important influence here. The asceticism of St Bernard had led to an emphasis on purity of form and architectural proportion. Transverse oblong bays and buttressed arches were employed by Cistercian architects before their adoption by the cathedral builders of the Île de France.

So not only had the experiments with new forms begun but the success of the pointed arch had been noted. The role of geometry now was to provide an intellectual foundation for further development. There has recently been great interest in examining the geometric proportions in twelfth-century buildings, both Romanesque and Gothic, by people who believe that Christian neo-Platonists designing churches had visualised certain proportions as symbolising theological ideas. There was a medieval preference for the ratio of 1:2, but in terms of two squares whose areas were proportioned as 1:2 rather than as two sides of a rectangle. This was said to be because the square symbolised the Godhead.[21]

Some modern scholars have discussed so-called 'sacred geometry' which relates mystical aspects of religion to geometric form. By measuring the proportions built into cathedrals they find it possible to suggest symbolic significance for some of the proportions employed. But there are aesthetic and structural grounds for identifying the increased emphasis on geometric form at this time. The place where Adelard's Euclid can readily be shown to have made an impact was Chartres, the most famous mathematical school in the Christian West.[22]

Soon after its appearance Adelard's Euclid became the basis for the study of geometry at Chartres. Only a few years later a disastrous fire required that the cathedral be rebuilt. Another fire then occurred, but the new west front had survived and is more than sufficient evidence of the basic understanding of geometry which underlay both the structural scheme and the decorative sculptures. Later developments carried forward what was initiated in 1145.

Of particular interest because of Adelard's earlier authorship of *De eodem* is the right-hand portal where the Mother of God is enthroned surrounded in the archivolts by statues personifying the Liberal Arts, each accompanied by a great classical master of the discipline. Geometry is accompanied by Euclid. The rendering of this theme is the first in monumental sculpture. In the statues in the jambs and in the groups of the Liberal Arts with their corresponding philosophers the basic proportion employed is the golden section. The tympanum is formed by an elaborate interrelationship between the equilateral triangle and a square. As one authority has said 'the felicitous thought of co-ordinating this theme with the figure of Mary as the Seat of Wisdom bespeaks the ultimate good and purpose of liberal studies at the Cathedral School.'[23]

Why is the golden section important? For one thing it has been taken throughout the ages to represent an almost perfect proportion. In Euclid's geometry it is referred to seventeen times. Not only in the design of the façade at Chartres is it in evidence, but also in the proportions of the north tower, where a pentagon, one matrix of the golden section, replaces an earlier square. A golden section is achieved by dividing a horizontal line with a vertical at a point where the parts of the horizontal line are in mean/extreme ratio. It is the visually satisfying proportion where the shorter section bears the same relationship to the longer that the longer has to the full length of the line. Its use at Chartres is an indication of a carefully calculated and harmonious theme based on mathematics rather than intuition.[24]

The new building at Chartres was begun in Adelard's lifetime so it is hoped that he was aware of the usefulness of his translation of Euclid beyond the confines of the classroom. It is thought also to have contributed to the development of Gothic architecture in the West Country of England. The movement presumably had its beginnings when Lalys was brought back from the Crusades to Glamorgan by Richard de Granville. Within the next fifty years there arose the most remarkable of the English schools of early Gothic, studied in detail by the late Sir Harold Brakspear and summarised by John Harvey:

> Perhaps beginning at Malmesbury Abbey about 1140, it culminated in the last

> years of the century in the choir of Lichfield Cathedral. The work of the school spread to Shrewsbury (St Mary's) in the north, to Glastonbury in the south, to Nuneaton in the east and to St David's Cathedral in the west; its greatest master was unquestionably the architect of Wells Cathedral . . . .
>
> The significance of this western school of design lies in its complete emancipation from survivals of the classical system, still seen in the Corinthianesque columns of the French work at Canterbury of 1175 . . . . [Wells] is not merely the earliest surviving building in England to display a fully emancipated Gothic aesthetic, it is the first anywhere to escape completely from the taste of Rome . . . .
>
> . . . Wells Cathedral in its original form stands as the noblest fruit of the first age of England. That age launched by the learned or at least well-tutored Henry I, has lasted until now upon the standard of that king's twelve-inch foot and upon the experimental science introduced by his great subject Adelard of Bath.[25]

Recent excavations at Wells have revealed 'a small but significant collection of fragments of the first half of the twelfth century' confirming documentary references which imply building works between 1142 and 1148. Bishop Robert (1136–66) gave Wells a new constitution and though he was still at Bath this foreshadowed the return of the bishopric to Wells. Bishop Robert was a friend of Henry of Blois, Bishop of Winchester (King Stephen's half-brother). The proximity of Bath to Bristol, Robert of Gloucester's sphere of influence (see below, p. 94), may have been a contributing factor. Henry of Blois was also Abbot of Glastonbury.[26]

Under Bishop Robert's successor, Bishop Reginald, these earlier works were abandoned and a new church in Early English style was begun *c.* 1180, a generation after Adelard's death. The choir at Wells was complete by 1190 and the transepts and east nave were finished by 1209. All these were built to a single design with slight changes of detail. Harvey describes the vaulting, with cross-arches and diagonal ribs, no ridges, but of high pitch and suave curvature as 'one of the most excellent in design of our early vaults'.[27]

The building proportions at Wells have been carefully examined and an analysis made. The author of this study finds that not only are there two interlocking schemes but that the methods of the master masons were both geometrical and arithmetical. Because Wells was built on a new site the masons were not impeded by an existing structure but could begin an entirely new design. The proportion which predominates is that which frequently occurs in medieval Gothic – that of $1:\sqrt{2}$, the relationship of the side of a square to its diagonal. It is the proportion which is at the basis of doubling the area of a square which has previously been mentioned, and related to the golden section. Master masons at this time were not in a position to use irrational numbers like $\sqrt{2}$. This would not have been a hindrance since the pattern could be achieved by laying out a scheme on a plaster floor. Alternatively the architect could use a drawing board. Known measurements which approximated the proportions could also have been used arithmetically. In the case of Wells the measurement was 29 ft (8.8 m) for the elevation of the triforium string, 41 ft (12.5 m) for the clerestory string and 14 ft 8 in (4.5 m) for the arcade capitals. The total height of the cathedral walls is 70 ft (21.3 m), 29 + 41 ft. The width of the building is 70 ft.

Working out the proportions one finds that 41 is the diagonal of a square with sides 29. It means that Wells was designed to be built to the square and that the central area between the transepts was a cube with sides of 70 ft.

Interwoven with this proportion there is another in the vaulting design: from floor to apex of the vaults is 66 ft (20.1 m), from floor to capital above the clerestory and supporting the vaulting is 44 ft (13.4 m), a different proportion for this height of 2:3. Thus two separate systems intertwine to produce the elevation, in both cases using figures which could be measured readily in terms of a standard English foot. They are close enough for the design of the vaulting to be incorporated within the overall scheme.

On the other hand the width of the nave and its relationship to the side aisles and to the bays as well as the design of the piers would indicate a purely geometric method of achieving the same proportions. The measurements involve complicated fractions but are quite straightforward as pure geometry. The width of the nave, 31 ft 8 in (9.6 m), is almost twice the nave bay lengths which are 16 ft (4.9 m). It could be thought that this reflects vaulting considerations since it is easy to calculate the span of the diagonal ribs of a rectangle whose sides are 1:2. However, bay lengths of the transept (14 ft 8 in, 4.5 m) and of the choir (17 ft 1 in, 5.2 m) are quite different and demonstrate that bay lengths had no part in the fixing of overall dimensions and vaulting considerations were independent of bay design. At Rheims and Chartres, which are similar to Wells in the use of pointed arches, the bay forms are no longer determined by the vaults and the designers have achieved a greater freedom and technological advance.

From the measurements it appears that the major dimensions at Wells were fixed by using established arithmetical approximations for the $1{:}\sqrt{2}$ proportion. Sometimes, as with the triforium and clerestory, the approximations appear to have coincided with an almost exact expression of the proportion; generally they did not. In the division of the nave width they do not appear to have been used, so that the resort here was to a purely geometric system. The total width of the arcade and aisle is $\sqrt{2}$ times the width of the aisle, a proportion which occurs in Norwich and may have occurred in Durham.

The point to be made is that whether the distances were intended to be simple divisons of whole feet or more complicated fractional figures, as would be produced by the geometrical method, they are not related to the arithmetical approximations which have hitherto produced all major measurements. Both systems were clearly available to masons and it may be that they chose between them or combined them in an arbitrary fashion.[28]

What is particularly interesting is that Wells was designed and built with no apparent association with the use of Gothic on the Continent. The new style at Canterbury was initiated by William of Sens. What connection the builders and designers of Somerset had with colleagues abroad or what mathematical skills they acquired at home is a matter of pure conjecture. The proportion which they chose was extremely popular and extremely versatile. Moreover, the pointed arches resemble those of the Seljuk Turks.

Medieval masons guarded their secrets; this analysis demonstrates the point. Harvey is of the opinion that a long period of architectural experimentation in the East before the Crusades provided solutions to problems of construction which were handed on orally or by demonstration when the Crusaders employed local builders. He believes that the 'secrets' involve more than the manipulation of geometric forms.[29] It is a point made indirectly by Talbot Rice describing a system of building in Isfahan where a square compartment was roofed by setting beams to form an octagon, then crossing and recrossing until the whole area was covered in brick vaulted ribs, which were used as frames on which the masonry of domes and vaults were supported. This method anticipated developments in the Gothic world by some centuries. Talbot Rice also makes the point that in the area dominated by the Seljuk Turks at the time of the Crusades there was secular building work involving fine stone masonry, pointed arches, elaborate voussoirs and defensive conceptions which were to follow in Romanesque and Gothic architecture a generation or so later.[30]

It was not very likely that a young man who wished to become a master mason would have pursued higher studies or that he acquired his geometry within a formal curriculum. Even Hugh of St Victor's *Practical Geometry* is a schoolman's textbook. In 1486 a German architect, Matthaus Roriczer, published a book entitled *On the Correct Building of Pinnacles* in which he demonstrated the art of taking an elevation from a plan. His method of designing pinnacles was also a method of designing other parts of a cathedral which is why it was important. Roriczer knew himself to be divulging a 'secret' and sought the permission of his bishop. By this time masons were organised in guilds which had agreed not to tell outsiders about the methods they used. But, as Jean Gimpel points out, two centuries earlier Villard de Honnecourt illustrated the geometric principles involved in his sketchbook. He did not consider that there was any secret about doubling the size of a square.

The clue to the mason's secret is that where measurements in Gothic buildings are not commensurable it must be assumed they were established by the geometric method. Masons believed strongly in the adaptability of what is called the $\pi/4$ triangle. This is an isosceles triangle with an apex of 45° derived from the regular octagon by drawing diagonals through the centre. This and the self-generating octagram based on the proportion of $1:\sqrt{2}$ occur with great frequency and are very easy to recognise. Originally proportion was used as a practical device because of lack of a yardstick; geometric methods were applied as a matter of expediency. But the recurrence of the same proportions automatically recalls an organic unit. This is the real relation between aesthetic design and the mason's skill in the creation of a beautiful building.[31]

One method by which medieval masons apparently handed on information to their successors was by leaving a geometric figure cut into the stonework. There are a number of pentagrams cut into the stonework of the cathedral at Nidaros in Norway which dates from the tenth century and an Archimedes spiral is inscribed on the inside face of one of the pillars of the triforium on the south side of the nave of Wells Cathedral.[32]

The proportioning of the arches in the Islamic world is basically similar to early Gothic. Arcs are drawn from the diagonals of squares to give ratios of $1:\sqrt{2}$. This simple method of establishing a commensurate system of proportions throughout a building was well known. The system had the advantage of deriving its ratios from the perfect square, a favoured shape in Islamic buildings century after century.[33]

These comments on the development of Gothic architecture should not be taken to imply that aspects of proportion were not well understood before the reintroduction of Euclid. Copies of Vitruvius were available in Charlemagne's time. In Book IV Vitruvius discusses the proportions of a religious building in terms of the rectangle formed by two identical squares (that is, the proportion of 1:2 – the golden section can be in fact derived from it). In Book I Vitruvius speaks of the importance of eurhythmy, of achieving symmetry between the individual elements of the building and the whole. He speaks of dynamic symmetry or symmetry of the second power. What this means is that there can be a relationship between two plane figures based on their combined proportions rather than on individual linear elements. For example the surfaces of two squares built on lengths proportional to $\sqrt{2}$ or $\sqrt{3}$ will be to each other as 2:3, that is, commensurable in the second power.[34]

Vitruvius was used in the Dark Ages more as a handbook on construction than a basis of design. Certainly his suggestions for temples were not considered suitable for the design of churches. Harvey believes that one of the best features of European Gothic was its emancipation from the theories of Vitruvius and quotes Enlart on how the designers maintained a human scale of values within their buildings which was one of their finest characteristics.[35]

In generalising one must restrict oneself to the belief that the availability of the complete Euclid was a tremendous stimulus to further development rather than an innovation. Adelard's writings reveal his interest in geometry, in proportion, and occasionally in building techniques, but not in cathedral or church design or in architecture as such. It is a pity that one cannot associate him directly with any one great building. This does not, however, diminish his contribution to scholarship; he certainly had a role in developing mathematical skills in the west of England. The fact that educated men now had this additional resource must have influenced those who employed designers and master craftsmen. They in turn were clever enough to develop building techniques based on sound theory.

Notes

1 Modern studies of Adelard's translation of Euclid's *Elements* were initiated by Marshall Clagett, 'The Medieval Translations from the Arabic of the *Elements* of Euclid with Special Emphasis on the Versions of Adelard of Bath', *Isis* 44 (1953), pp. 16–42. More recently other scholars have elaborated on his findings, some of which were confirmed and others refuted. Richard Lorch, 'Some Remarks on the Arabic-Latin Euclid', *WIST* XIV, pp. 45–54; Menso Folkerts, 'Adelard's Versions of Euclid's *Elements*', *WIST* XIV, pp. 55–68; H.L.L. Busard, *The First Latin Translation of Euclid's 'Elements' Commonly Ascribed to Adelard of Bath*, Pontifical Institute of Mediaeval Studies, Studies and Texts 64 (Toronto, 1983); see also review of Busard by Charles Burnett, *Archives Internationales d'Histoire des Sciences* 35 (1985), pp. 475–80; *WIST* XIV, Catalogue 15, pp. 170, 171.

2 Adelard of Bath, *De opere astrolapsus*, Cambridge, Fitzwilliam Museum, McClean MS 165, f. 84v.

3 See above, ch. 2; Folkerts, *WIST* XIV, pp. 60, 61.

4 Lorch, *WIST* XIV, p. 45.

5 *WIST* XIV, Catalogue 81, p. 186.

6 Folkerts, *WIST* XIV, p. 59, believes Adelard I may have been the work of another translator and that Adelard II is the translation referred to in *De opere astrolapsus.*

7 Folkerts, *WIST* XIV, p. 59, refers to fifty-six extant manuscripts; Clagett 1953, p. 21, gives fifty-five.

8 Clagett 1953, p. 21. For a possible identification of 'Ocreatus' as a member of the Hosatus family see Charles Burnett, 'Ocreatus' in M. Folkerts and J.P. Hogendijk (eds), *Vestigia Mathematica* (Amsterdam, 1993) pp. 69–78.

9 Clagett 1953, pp. 19, 23.

10 Busard 1983, *passim*; Clagett 1953, pp. 26, 27; Folkerts, *WIST* XIV, pp. 55–62, notes.

11 Alistair Crombie, 'Science', ch. 18, in A.L. Poole (ed.), *Medieval England*, 2 vols (Oxford, 1958), vol. 2, p. 580:

> It is scarcely possible to exaggerate the importance of Adelard's translation of Euclid's *Elements*. Before it became available to them the Latin mathematicians and natural philosophers had only the conclusions of some of Euclid's theorems and perhaps the proofs of one or two. Adelard's translation introduced them to the full conception of Greek axiomatic method and provided a model for their scientific thinking.

John Murdoch, 'The Medieval Euclid: Salient Aspects of the Translations of the *Elements* by Adelard of Bath and Campanus of Novara', *XIIè Congrès international d'histoire des sciences, Colloques (Revue de synthèse)*, 3ème série, nos 49–52 (Paris, 1968), pp. 67–94; 'Euclid: Transmission of the *Elements*', in *DSB*, vol. 4 (1971), pp. 437–59. Murdoch states (pp. 445, 446) that by far the most important share of the medieval Arabic/Latin Euclid belongs to Adelard of Bath.

12 See above, ch. 4, p. 40, n. 16.

13 Ivor Bulmer-Thomas, 'Euclid and Medieval Architecture', *The Archaeological Journal* 136 (1979), pp. 135–50, (pp. 141–7). For the Masonic MSS: John H. Harvey, *The Mediaeval Architect* (London, 1972), gives full text of the Cooke MS with modernised spelling (pp. 191–202) and the variant sections of the Regius poem (pp. 202–7).

14 Stephen K. Victor, *Practical Geometry in the High Middle Ages*, American Philosophical Society Memoir 134 (1979), pp. 7–18, 43, 44, quotes John Murdoch as stressing Adelard's effort to relate the whole geometry of commensurable magnitudes to the science of practical measure.

15 Lon R. Shelby, 'Geometrical Knowledge of Medieval Master Masons', *Speculum* 47 (1972), pp. 395–421; Paul Frankl, 'The Secret of the Medieval Masons', *Art Bulletin of New York* 21 (1945), pp. 48, 60.

16 John H. Harvey, 'The Origins of Gothic Architecture', *Antiquaries Journal* 48 (1968), pp. 87–99. David Talbot Rice, *Islamic Art* (London, 1965), pp. 59, 86–9, 165–8.

17 Harvey 1972, pp. 29, 30, 94–7; see also K.J. Conant, *Carolingian and Romanesque Architecture from 800–1200* (Harmondsworth, 1959), pp. 223, 256, 290, 291.

18 Harvey 1968, pp. 91–4; E. Jervoise, *The Ancient Bridges of Mid and Eastern England* (London, 1932), pp. 142, 143.

19 Martin S. Briggs, 'Building Construction', in C. Singer *et al.* (eds), *A History of Technology* vol. 2, *The Mediterranean Civilization and the Middle Ages* (London, 1956), pp. 397–446 (435).

20 E.H. Gombrich, *The Story of Art* (London, 1957), pp. 123–32.

21 Mâle 1948, pp. 321–3; Otto von Simson, *The Gothic Cathedral* (London, 1956), *passim*, mathematical symbolism pp. 35–55, Appendix by E. Levy, pp. 235–67.

22 George Lesser, *Gothic Cathedrals and Sacred Geometry*, 3 vols, vol. 3, *Chartres* (London, 1952).

23 Von Simson, p. 153; F.M. Lund, *Ad Quadratum*, 2 vols (London, 1921); Mahla Ghyka, *Geometrical Composition and Design* (London, 1952).

24 John James, *Chartres, the Masons Who Built a Legend* (London, 1982), p. 114; Lund, p. 134; von Simson, p. 208.

25 John H. Harvey, 'Wells and Early Gothic', *FWCR* (1978), pp. 19–24; Harvey 1972, pp. 30, 94–112.

26 W. Rodwell, 'Excavations and Discoveries', *FWCR* (1980); *The Archaeology of the English Church* (London, 1981); K. Potter (ed.), *Gesta Stephani* (Oxford, 1976), with new introduction and notes by R.H.C. Davis, pp. xviii-xxxviii. Davis suggests that Bishop Robert is the anonymous author. Adelard was presumably resident in Bath during the civil war.

27 John H. Harvey, *The Cathedrals of England and Wales* (London, 1974), pp.119, 120.

28 Barrie Singleton, 'Proportions in the Design of the Early Gothic Cathedral at Wells', *British Architectural Association Conference Transactions: Medieval Art and Architecture at Wells Cathedral* (1978). I am grateful to the late Mr L.S. Colchester for bringing this report to my attention. See also L.S. Colchester, 'Dimensions and Design', *FWCR* (1981), pp. 15–19; L.S. Colchester and John H. Harvey, 'Wells Cathedral', *The Archaeological Journal* 131 (1974), pp. 200–14 and fig. 2, drawing by John H. Harvey of plaster floor used for trials of geometric designs based on photographs in the National Monuments Record.

29 John H. Harvey, 'Geometry and Gothic Design', *Transactions of the Ancient Monuments Society* 30 (1986), pp. 43–52.

30 David Talbot Rice, *Islamic Art* (London, 1965), pp. 59, 60, 87–9, 165–8.

31 Frankl, pp. 45–60; Shelby, pp. 397,398. See also Jean Gimpel, trans. Teresa Waugh, *The Cathedral Builders* (Salisbury, 1983, new edn 1988), pp. 75–97, trans. Carl F. Barnes, Jr (New York, London, 1961), pp. 107–45.

32 Lund, pp. 129, 186. The Archimedes spiral inscribed on a pillar in the nave of Wells Cathedral has been recorded by the Royal Commission on the Historical Monuments of England for the National Monuments Record, Ref. no BB70/7430.

33 Ronald Lewcock, 'Architects, Craftsmen and Builders: Materials and Techniques', in George Michell (ed.), *Architecture of the Islamic World* (London, 1978), p. 132.

34 Ghyka, pp. 14, 15.

35 Harvey 1986, p. 43.

CHAPTER 8

Adelard and al-Khwārizmī's *Zīj*

Adelard's translation of the *Zīj*, or astronomical tables, of al-Khwārizmī ranks in importance with his Euclid. There are those who consider it to have been the more significant contribution to Western scientific thought.[1] The role of al-Khwārizmī as mathematician and astronomer was so important that it was a big step when the data he had collated and explained became known in Christian Europe. His works were widely disseminated in the Arab world, in Spain and Morocco as well as the East. It was after the reconquest by the Christians of Toledo in 1085 that opportunities to study Arab astronomy became readily available. Adelard was only one of several translators, but again his work seems to have made the greatest impact.[2]

Al-Khwārizmī had been a key scientist in the ninth-century House of Wisdom in Baghdad founded by Caliph al-Ma'mūn. But the groundwork of his mathematics had been laid a century before when a man from India arrived at the court of al-Mansūr. He was skilled in the 'calculus of the stars', and brought with him a Sanskrit astronomical text known as the *Sindhind* (*sidhānta*), and possessed methods for solving equations based on sines calculated for every half degree, also methods for computing eclipses. Al-Mansūr ordered the book in which all this was contained to be translated into Arabic so that it might serve as a foundation for computing the revolutions of the planets. This was done by al-Fāzāri whose work the Arabs called the *Great Sindhind*. An abstract of this was made by al-Khwārizmī who used it to prepare his tables a century later.[3]

Al-Ma'mūn and the House of Wisdom promoted not only al-Khwārizmī's work but called on other learned men to translate Ptolemy's *Almagest* from the Greek. This fact also influenced the development of Arab astronomy. The *Almagest*, compiled in the second century AD by Claudius Ptolemy in Alexandria, was a comprehensive summary of available Greek and Egyptian theory. It served as the basic textbook of astronomy until Copernicus. Under al-Ma'mūn experiments were conducted both at Baghdad and Damascus to verify and correct Ptolemy's tables.[4]

Ptolemy used the Babylonian method of dividing a complete circle into sixty parts which were then subdivided twice more. The unit of length in which a chord was measured was obtained from the radius of its own circle. The method was a geometric one. The Arabs altered the calculus of chords used by Ptolemy into the calculus of sines and trigonometry. They added other functions to the early tables. When the proportion could later be given as a decimal fraction trigonometry became even more useful.[5]

Adelard was acquainted with the ideas of simple astronomy before his journey to Syria. He shows this in *De eodem* when he presents his description of Astronomy as one of the Liberal Arts.[6] Medieval astronomers pictured the celestial sphere divided by great circles in much the same way as we do. The equator divides northern hemisphere from southern, and parallels of latitude are measured in relation to it. The colures are two great circles at right angles to each other crossing the poles and intersecting the great circle of the ecliptic, which traces the apparent course of the sun throughout the year. On the equator at east and west the points of intersection are the equinoxes, where day and night are equal in length. The north/south colure passes through the solstices where the sun reaches its most northerly point in summer and its most southerly in winter. The ecliptic is divided into the twelve signs of the zodiac, each of 30°, and its plane is tilted at an angle of approximately 23°27′ to the equator. Ptolemy's estimate of the ecliptic was appreciably greater, 23°51′, and in his view the sun was moving around the earth instead of being the centre of the solar system.

Co-ordinates pinpointing positions on the celestial sphere can be established for heavenly bodies in relation to these intersecting great circles, the angles which they make and the arcs which they produce. Modern astronomers use the term 'right ascension' and 'declination' for measuring the apparent positions of celestial bodies in relation to the celestial equator and the great circle through the equinoxes. Ptolemy's system determined a star's latitude and longitude in relation to the ecliptic.[7]

Adelard recognised that it was difficult for people to grasp the significance of the obliquity of the ecliptic. In *Quaestiones naturales* he has his nephew comment 'Surely it would be better if the universe were arranged symmetrically', to which Adelard replies 'If the zodiac and stars followed a straight line we would be without winter and summer .... Let us hear no more of your evil suggestion of a perpetual spring'.[8]

For non-astronomers the essential point to grasp is that intersecting great circles cut off triangles on the surface of the sphere. Astronomers have an imaginary concept of these in the sky as geographers have of the equator on the surface of the earth. The arcs of these triangles form chords which in their turn can be used for calculations similar to those of triangles on a plane surface. The Ptolemaic method derives from Greek theories. The sine method originated in Hindu astronomy. The two methods are easily interchangeable once the basic principle is understood.[9]

In all cases where men in the Middle Ages used a *zij*, we use an ephemeris – both are sets of tables in which information is recorded so that an observer can calculate the movements of heavenly bodies not just from day to day but from minute to minute. This type of table has a very long history. In the centuries immediately preceding Islam there was at least one astronomical handbook and probably more were compiled and used in Sassanian Iran showing strong Hindu influence. According to the authority E.S. Kennedy:

> In the eight centuries beginning with AD 700 well over 100 different *zijes* were produced. Of these more than twenty were based partially on observations

made by their respective authors. Others differed from the prototypes only to the extent that mean motion tables were recomputed for a different epoch and calendar. In the majority of these the theory is that of Ptolemy with some improvement of parameters. The main Islamic contributions were in trigonometric computational and observational technique.

A minority of these *zijes* were based on Hindu or pre-Islamic Iranian theory – the only one preserved is the al-Khwārizmī *zij* which has:

a) Planetary equations from the Sassanian Shah *zij*

b) Planetary latitudes not presently identifiable but non-Ptolemaic and probably Hindu

c) Planetary stations from the Almagest and hence irreconcilable with a) and b) above

d) Two values for the length of the sidereal year, one a common Hindu parameter used in the Brahma Siddhanta, the Siddhanta Siromani and the Arabic Sindhind; the other a Persian value.[10]

Al-Khwārizmī's original *Zij* no longer exists and Adelard's Latin translation is therefore tremendously important. It is based on a revision of Maslama (fl. AD 1000) and was designed to be used in Spain at Cordova. Neugebauer makes the point that as a result of Adelard's translation England and France were influenced by a work which reflected a mixture of Greek and Hindu elements at a time when these had been outdated by eastern Islamic work.[11] The *Zij* is of particular interest because so much of the material is non-Ptolemaic. This comes as a surprise to those who have looked first at Adelard's Introduction where he makes the point that unless one has made an effort to understand the theories of Ptolemy one might think al-Khwārizmī's tables were merely constructed by using certain formulae that had been handed down. Knowledge of the *Almagest* indicates that the calculations are based on sound theory. In the Introduction (Neugebauer's translation) Adelard states, '. . . if he [the reader] has been well trained in the use of these rules – and has also made a trial of Ptolemy's *Almagest* – then he will not doubt that all that follows here originates from necessity.'[12]

No Latin version of the *Almagest* was available to Adelard but he certainly knew about it. Suter suggests that al-Khwārizmī was acquainted with the *Almagest*.[13] References in the Arabic text might have been later additions.

Continuing his Introduction Adelard first explains the concept of Arin, or Arim, the zero meridian of Hindu astronomy which al-Khwārizmī employs. To understand what is meant by Arin it is essential to appreciate the role which the Greenwich meridian plays in all modern calculations of time. Arab calculations were made in terms of a meridian at Arin in exactly the same way, first by calculating one's own position on the earth's surface in relation to this meridian, or one's longitude, and then calculating the local distance from Arin for each tabulated figure in which one was interested. The observatory at Arin is assumed to be located at the midpoint of the hemisphere which extends from the Ocean in the West to the Ocean in the East and from pole to pole. (Its site in modern terms is that of Ujjain in India, 76°E. of Greenwich and 23°N. latitude.)

The significance of the twenty-third parallel of latitude for estimating where the sun will be directly overhead at the solstice has previously been mentioned (Chapter 4, p. 34, p. 40, n. 10). Although Arin is regarded as a zero meridian in modern terms, in the Middle Ages it was thought of as a point not only midway from east to west but also from north to south. Arab astronomers were under the impression that Arin or Ujjain was on the equator but this was not so. An imaginary great circle through Ujjain connected Lanka with the semi-legendary Mount Meru behind the mountains of the Himalayas, the source of India's rivers. It was Lanka that Hindu astronomers considered to be the centre of the earth and without latitude. Arin was designated by Ptolemy as Ozēnē (hence Arabic Uzayn, regularly corrupted to Arīn, and further characterised by the Arabs as *qubbat al-ard*, 'the cupola of the earth' or 'world summit'). This idea was formulated by the Arab geographer Ibn Rusta writing *c.* 290/903. Ptolemy, however, did not measure longitude from Arin but from the westernmost point known to him, the Canary Islands. Al-Bīrūnī thought this was a better method than al-Khwārizmī's, but Ptolemy was 20° out in his measurement of the length of the Mediterranean. Al-Khwārizmī corrected 10° of this.

Inaccurate estimates of the longitudes of cities where tables were going to be used caused many difficulties.[14] A number of examples can be given. The distance between Baghdad and Cordova was thought to be 63°. Al-Battānī, another Arab astronomer, gives for Baghdad the longitude of 80° from the westernmost point, while al-Khwārizmī gives 70°. This means Cordova is either 7° or 17° from that same point. Elsewhere Cordova's longitude is recorded as 9°9′, 8°40′ or 18°40′. These variations show how impossible the compilation of accurate tables was until problems of determining longitude were solved with the chronometer.[15] Nevertheless, Adelard gives the impression that he understands the principles and can help others to do so.

When the first point of Aries, representing the start of the vernal equinox, was on the meridian due south at Arin it provided the starting-point for measuring the angular difference in time or degrees of longitude for the appearance of celestial bodies on the meridian of an observer, but only so long as he knew his distance in longitude from Arin. The difference in time then as now was one hour for 15°. If longitude was incorrectly assessed it obviously affected the usefulness of tables, unless they were used where the observations were originally made. Al-Khwārizmī's tables based on Arin recorded positions of the seven planets and of the ascending nodes of the moon reckoned from noon of day 4 (Wednesday) to noon of day 5 (Thursday), and all other days were similarly reckoned from noon to noon. Noon of day 4 was the beginning of the Hijra and also of the month of Muharram, and so the *zij* takes this for its beginning.

Not only were the celestial phenomena recorded in terms of the meridian at Arin but there was also the question of correlating the calendar dates since different nations used different calendars. This was the first problem to which Adelard addressed himself. There has been very careful scholarly editing of the *Zij*. I use Neugebauer's translation for my quotations. The reason why Adelard's work is thought to have been produced in 1126 is because this year is given as equivalent to the Arabic year 520 and used as an example of how the Arabic year made up of lunar months (354 days) and the Roman year (365 days) could be related to each other. The mathematical procedures are not difficult to follow.

Although the preliminary remarks in one manuscript refer to a number of tables, only one table has been found which can be associated with the introductory text, that giving thirty-year groups of Arabic years which are equated with *anni Alexandri*, that is, years of the Seleucid era, not the Christian one. This is Table 3.[16]

In all the computations 'months' are counted as thirty days, whereas in the Arabic calendar alternate months were only twenty-nine days. This means that if twelve months are reduced to one Arabic year, six days have to be added. Similarly in the counting of excess months, one day has to be added for every two months. In his Introduction Adelard also explains that for the Roman solar year five and a half days have to be subtracted to achieve twelve months of thirty days. Adelard establishes that from the beginning of the Christian era to the Hijra was 621 years, six months and fifteen days:

> Since this book is intended to be useful for different nations, of which not all follow the same time (reckoning), a table is given in the following, which contains the times of kings as well as of events, such that, when these rules are given and noted according to the time of the Hijra, time and years can be reduced to any definition of different nations.
>
> ... Since this book, as mentioned before, determines the positions of the planets and the ascending nodes according to Arabic years and months and since we are concerned about its usefulness for their Latin (forms), we note first of all the table given below, by means of which one can obtain from any Roman years and months and days, beginning with the very year in which this book was translated into our language, for unlimited time, the correspondence with Arabic years, months, and days.[17]

A quick rough calculation will show how Adelard arrives at the figure 520 = 1126, but for astronomical calculations not only do years have to be changed into months and months into days but each day is subdivided into twenty-four hours, hours into sixty *dakaicas* (or degrees or minutes) and each *dakaica* into sixty *zenias* or seconds. He then proceeds to show in Chapters 6 and 7 how the date and time of the Arabic year can be used with other tables to find the mean position of the planets first for noon at Arin and then for noon in any region.

It is worth noting that the use of a symbol for zero is an accepted procedure in relation to tables. Here, for example, in Chapter 7, 'How to Find the *Elwacat*, or Mean Place, of the Planets', Adelard instructs the reader with regard to tables 4 to 20:

> You enter for any (given) moment in the tables called *elwacat* of the planets for any particular planet with accumulated years then single years, then with months, days, hours and minutes all of them completed – the total of the (numbers) found there you reduce beginning with the smallest units, i.e. seconds.
>
> If 12 signs accumulate by this reduction, you ignore them and note down only the remainder of signs. Otherwise you write down instead of 12 the symbol of the *cifra* τ. Similarly for the other units. The whole remainder obtained in this way will be the *elwacat* of the given planet for noon of the locality of Arin for which this instruction was made.

The *cifra* τ is merely a terminus, not a real zero in our sense. In Chapter 8, 'How to find the Place of the Sun', Adelard explains the symbol of *cifra* with the words, '*id est nullus*', or in MS O *nihilum*. In other words the symbol represented nought, nothing. Richard Lemay has given this explanation:

> The presence of a zero in *ghubār* numbers came about because when a circle, like the zodiac, was traversed from beginning to end, and the progression of degrees, minutes, seconds and so on reached the limit of the category (thirty degrees for a sign of the zodiac, a total of sixty minutes in a degree, sixty seconds in a minute and so on) and before the next series started, the exact value in the tables was null and was so indicated by τ. In many twelfth-century astronomical tables where this value had to be indicated the circle was avoided because it often stood already for 8 (octo).[18]

Returning to the concluding section of Chapter 7, an underlying interest in geography is revealed but little use seems to have been made of it. A method is described of measuring degrees of celestial latitude, and relating them to earthly equivalents:

> Also note that according to the Chaldeans, 4,000 camel paces make one mile, and that 33⅓ miles on the earth (correspond to) ½ degree on the heaven, such that the whole circumference of the earth contains 24,000 miles. The reason for it is: if one travels from any place directly south, then, after going 66⅔ miles, a star observed at the first place, will then be seen (at the second place) one degree higher, (if observed) at the same minute of the hour. Consequently 1 degree and ½ make 100 miles, thus 15 (degrees) 1,000 miles, 1 sign 2,000 (miles) thus 12 (signs) 24,000 (miles).[19]

Al-Khwārizmī has made or recorded someone else's calculations for the miles involved in one degree of latitude and is aware that it is possible to use celestial observation for calculations related to the earth's surface. The 'twelve signs' are the signs of the zodiac, and if each represents 30° it would indicate that each time the sun's position in the sky is represented by a new sign the equivalent earthly journey is 2,000 miles. A full cycle of all twelve signs is 24,000, approximately the earth's circumference. Signs of the zodiac were very familiar to men and women of Adelard's day. They were often represented in sculpture in Romanesque churches.

Bearing in mind that more than two centuries had elapsed since al-Khwārizmī's tables were first produced and that two different methods of calculating the length of the year were involved, it is not surprising that Adelard's text is further from the original version than was first realised. The discovery of another translation of al-Khwārizmī based on the same original source but with a correlation in terms of 1115–16 raised further questions. The second translation was by Petrus Alfonsi, the converted Jewish physician at Henry I's court. There are three texts of his work. Petrus could have brought the tables with him from Spain. He was well versed in astronomy as is shown by an elaborate explanation of Arin in his *Dialogi cum Judaeo* about his conversion to Christianity. He must have been a man of some importance because Alfonso I of Aragon stood godfather for him and gave him his name. Petrus is known to have wanted to write an astronomical treatise. While in England he assisted Walcher of Malvern with his astronomical observations. Because of the dates there seem to be sound reasons for suggesting a connection between his work and Adelard's. It is even suggested that Petrus could have had a major influence on Adelard's career and that for his understanding of astronomy Adelard owed much to Petrus's instruction.[20]

Whether or not there was collaboration between the two cannot be established. Adelard does not refer to Petrus when he speaks of his own translation of the *Zij*. Mercier suggests that there was a common source for both translations that no longer exists.[21]

After his careful study of the texts available to him Neugebauer came to the conclusion that Adelard's translation is probably closer to the original than the rather haphazard collation presented by Petrus. Neugebauer remarks sadly: 'No other conclusion is possible but that the majority of medieval astronomical manuscripts from India to England owe their existence much more to the prestige they gave their owners than to a real desire to master the underlying astronomical theory.'[22] Neugebauer makes this comment because of the number of discrepancies and errors in the texts. It seems that although the techniques were available they were not effectively utilised. For example, despite knowing how many miles were equal to a degree of longitude or latitude, no attempt was made to collate the information.

Chapters 7 to 17 which give al-Khwārizmī's instructions for using the tables are included by both Adelard and Petrus. In his detailed comments on the texts Neugebauer explains where the methods differ from those of Ptolemy. In Chapter 3 Adelard speaks of philosophers investigating the orbits of the planets and the circumference of the orbits – this might be a formulation influenced by Hindu tradition.[23] In Chapter 10 al-Khwārizmī's procedure for finding the true longitude of a planet is based on Hindu methods. The underlying model was not influenced by Ptolemy's great innovation of a motion regulated by an eccentric equant.[24] The description of lunar motion with only one inequality 'makes it evident that we are actually dealing with a type of astronomy uninfluenced by the *Almagest*'.[25] The tables cannot be used by modern scholars unless they understand the implications of all the variations. Incorrect results are obtained by those unwittingly combining different types of years, for example, in their chronological computations. Translating the *Almagest* from the Greek (*c.* 1150), Hermann of Carinthia comments on the inadequacy of the tables, as warned by al-Bāttāni, and substitutes some of his own.[26]

Important to any general survey of the history of science is the fact that Adelard's translation made al-Khwārizmī's trigonometric tables available in Europe, and he included explanations which stressed the importance of the sine. He may not have realised himself how many varied theories of planetary motion had become involved. He seems to have thought the work was closely allied to the *Almagest*. He says at the end of Chapter 23:

> Whoever desires to obtain skill in the field of astronomy should grasp this treatise with the application of his whole mind. On this, indeed, depends rightly not only the addition and subtraction of the correction [Neugebauer's note, i.e. the determination of the positive and negative corrections (equation of centre etc)] but also the determination of longitudes themselves. If anyone questions why to given arguments such sines be assigned, and conversely, he

should know that he can get the reason for it from the Almagest of Ptolemy. According to Neugebauer Adelard is probably referring to sines of longitude of celestial bodies and his statement is more appropriate for Hindu treatises than for his present work since the corrections were tabulated and ready for use.[27]

Adelard did not use the term *sinus*; this was introduced by Robert of Chester when he revised Adelard's text. This should not be taken to indicate any lack of understanding. According to Mercier, the term *sinus* was introduced on the assumption of an Arabic word meaning *al-jaib* ('pocket' or 'hollow'), while in fact it is only a misreading of an Arabic transcription of the Sanskrit *jīva* ('bowstring'), which was a good metaphor for a chord in a circle. There was not a suitable Latin word. Or rather if they had understood the meaning behind the disguised transcription of the Sanskrit it would be chord, *basis*. The use of a Latin word meaning 'pocket' or 'bay' actually stands in the way of understanding. It is interesting to observe Copernicus' studious avoidance of *sinus*, in his effort to be faithful to Greek and humanist dispositions. He always used *subtensa*.[28]

The *Zīj* contained a table of tangent functions as well as that of sines. Adelard was already well acquainted with the principles involved when the shadow function was used to convert altitude measurements with a gnomon into angles. He would have had no difficulty with the further application of the proportions applicable to right-angled triangles.

It is felt that because Western astronomy developed from Spanish sources it did not have the higher level and inner consistency of Byzantine and Islamic astronomy, a point which one might illustrate by reference to the experiment performed by al-Bīrūnī mentioned above (Chapter 4, p.40, n. 10). Al-Bīrūnī's experiment was based on a theory of al-Ma'mūn's astronomers that the circumference of the earth could be measured by taking the dip measure of the sun from a mountain peak rather than a measurement of the changing altitude of the sun.

Al-Bīrūnī found a suitable site at the top of a steep cliff with a sheer drop to a vast expanse of plain as far as the horizon – an alternative to the sea. He could take an accurate measurement of the height of the cliff and of the angle of dip. Using the dip to help him calculate the sine of the angle he was able to provide a measurement for the mileage involved in a degree of arc, and thus used a different method of measuring the earth's circumference trigonometrically.[29]

It is quite apparent, however, that the main concern of those using the tables was for astrology rather than for scientific astronomy. There were twenty-six folios devoted to Chapter 37, 'How to find the Aspect of the Stars in Sextile, Quartile or Trine', and its accompanying tables were numbers 91 to 114. For Suter this illustrated better than any other fact that the scientific work of antiquity and the Middle Ages in the area of astronomy was not an end in itself but a means to meeting the demands of astrology.[30]

The translation of another important work by al-Khwārizmī has been attributed to Magister A, thought at one time to have been Adelard of Bath. Now this is less certain. According to Lemay the translation of the tables overshadowed al-Khwārizmī's arithmetic, a treatise on how to calculate with Hindu numbers using a symbol for zero.[31] The arithmetical procedures described came to be called the 'algorism' after al-Khwārizmī.

The attribution to Magister A occurs in only one manuscript of the text – there are eight. It is a work in five books on the four parts of the quadrivium. One of these books bears a close relationship to the work of Petrus Alfonsi, a fact which gave rise to the suggestion that he, rather than Adelard, might have been reponsible for the translation. This is no longer considered likely, although he could have been the author of that part of the work without being the translator of the al-Khwārizmī arithmetic. The *Liber ysagogarum Alchorismi* 'gives the impression of being a rather unintelligent collection of excerpts from other texts'. Opinions differ as to Adelard's role. The fact that the text includes references to Euclid's *Elements* which were not part of the Boethian tradition has strengthened but not proved the case for Adelard as Magister A.[32]

The Hindu arithmetical procedures described in this text are based on a decimal place system with nine numerals combined with zero. The original methods of computation called *algorismus*, or algorism, were very similar to abacus techniques. Those who used them employed the symbol for zero instead of an empty column. Considerable excitement is shown at the possibility of writing out a series of digits and then expressing the full Latin text of a phenomenally long number.[33]

Interest in al-Khwārizmī's work was strong in Toledo after this city became the capital of Castile and a focal point for the transmission of Arabic ideas by way of Latin translations. *Ghubār* (or 'dust') numbers were formed by using a stylus or one of the points of a pair of geometric compasses to draw the numerals in a tray of wet sand. Alternatively they were inscribed on buttons or *apices* (see Chapter 3). Lemay considers that the form they took was related to al-Khwārizmī's method of employing them. Adelard refers to the use of a sand table in an edition of his Euclid, that is, Adelard III, and also in *Regule abaci*. He was undoubtedly aware of the usefulness of the number forms; nevertheless, he still used roman numerals in his works. Roman numerals continued for some time as did use of the abacus: scribes of scientific manuscripts in Latin Europe continued depending upon and refining the roman system. Only rare attempts were made during the twelfth century to introduce the full *ghubār* system into scientific manuscripts except when the subject was the algorism. A symbol for zero did not exist in the early *ghubār* but came into being because of its usefulness in astronomical tables. Despite its obvious convenience, certainly to our eyes, the transition to the new system was extremely slow. This would await the introduction of paper as something handy to write on. Unquestionably, however, the transition from the abacus to the algorism as a method of calculation was of particular importance in the development of mathematical ideas in the West.[34]

Sines also were computed in Ptolemaic chords rather than by trigonometry. Even in Arab countries new ideas were not adopted universally or rapidly. Al-Khazini in Marw was using Ptolemaic methods for the Sanjari *Zij* produced in 1115–16.

It is interesting, in connection with the influence which Adelard may have had in promoting new ideas in the west of England, that while arabic numerals were used by masons in Wells for identifying sculptured figures during the thirteenth century, roman numerals were employed in cathedral accounts for a further 400 years.[35]

Notes

1 Gibson, *WIST* XIV, p. 15.

2 Adelard of Bath, *Tables of al-Khwārizmī*, *WIST* XIV, Catalogue 34, p. 178; H. Suter *et al.*, *Die astronomischen Tafeln des Muhammed ibn Mūsā al-Khwārizmī in der Bearbeitung des Maslama ibn Ahmed al-Madjrītī und der latein Übersetzung des Athelard von Bath auf Grund der Vorarbeiten von A. Bjornbo und R.O. Besthorn*, Det Kongelige Danske Videnskabernes Selskab, histor.-filosof. Skrifter, 7 Raekke, Afd. 3 (Copenhagen, 1914); trans. and further commentary by O. Neugebauer, *The Astronomical Tables of al-Khwārizmī*, Det Kongelige Danske Videnskabernes Selskab, histor.-filosof. Skrifter, 4,2 (Copenhagen, 1962); Raymond Mercier, 'Astronomical Tables in the Twelfth Century', *WIST* XIV, pp. 87–118 . Mercier brings up to date the extensive research by earlier scholars on interrelationships among the extant manuscripts. Adelard himself refers to the *ezich* (*zij*) he translated from Arabic into Latin in *De opere astrolapsus*, McClean MS 165, f. 83v.

3 G.J. Toomer, 'Al-Khwārizmī', *DSB*, vol. 7 (1973), pp. 358–65; J.L.E. Dreyer, *History of the Planetary Systems from Thales to Kepler* (Cambridge, 1906), p. 244.

4 A. Sayili, *The Observatory in Islam*, Publications of the Turkish Historical Society, series 7, 38 (1960), Ankara, pp. 80, 175.

5 O. Neugebauer, *A History of Ancient Mathematical Astronomy* 3 parts (Berlin, 1975), part 1, Introduction, pp. 1–17 (9), book 1A, *Spherical Astronomy*, pp. 21–52.

6 *De eodem*, pp. 31, 32. See above, ch. 2, pp. 16–17 for description of 'Astronomy', trans. Theodore Wedel, *The Mediaeval Attitude towards Astrology particularly in England*, Yale Studies in English 60 (New Haven, Conn., 1920), pp. 49, 50.

7 Claudius Ptolemy, *The Almagest*, trans. G. Taliaferro, *Great Books of the Western World*, vol. 16 (Chicago, London, 1963); another translation annotated by G.J. Toomer (London, 1984). For a brief description of Ptolemaic system see *The New Columbia Encyclopedia* (New York, 1975), p. 2238.

8 *Quaestiones*, Müller, ch. lxxi, pp. 63, 64; Gollancz, p. 155.

9 Neugebauer 1975 book 1A, p. 21; Mercier, pp. 88, 89. John D. North explains mathematical background in *Richard of Wallingford*, 3 vols (London, 1976), vol. 3, pp. 142–54.

10 E.S. Kennedy, 'A Survey of Islamic Astronomical Tables', *TAPS*, N.S. 46 (Philadelphia, 1956), part 2, 14, pp. 128, 148, 173.

11 Neugebauer 1975, part 1, Introduction, p. 12.

12 Adelard's references to Ptolemy's *Almagest*: Neugebauer 1962, section I (Latin version as originally transcribed by Suter) p.2; section II (Neugebauer's trans.), p. 10.

13 Suter 1914, p. 33.

14 J.K. Wright, 'Notes on the knowledge of latitudes and longitudes in the middle ages', *Isis* 5, part 1 (1923), pp. 75–98; D.M. Dunlop, *Arab Civilization to A.D.1500* (London, 1971), pp. 154–7. Dunlop gives a detailed explanation of the background of Arin in relation to Indian theory. Neugebauer 1962, II, pp.10,11.

15 Neugebauer 1962, II, p. 111, n. 5.

16 Choice of year 1126, Neugebauer 1962, I, p. 1; II, p.9; *anni Alexandri*, I, p. 5; II, pp. 9, 10, 11. Mercier (p. 89) considers two primary texts of al-Khwārizmī's *Zij* to be C and O in which this date occurs. Others are incomplete, or attributed to Petrus Alfonsi. N is Robert of Chester's revision of Adelard. In ch. 2, O leaves out reference to tables which are missing. Mercier (pp. 97–100) gives a brief analysis of relations between manuscripts.

17 Adelard's method of converting calendar: Neugebauer 1962, I, pp. 4, 5; II, pp. 12, 15.

18 Use of zero: Neugebauer 1962, I, pp. 8, 10; II, pp. 18, 20. Neugebauer refers to the use of zero in Greek astronomy, which preceded by centuries the beginning of Hindu astronomy, in *Exact Science in Antiquity*, 2nd edn (Providence, Rhode Island, 1957), p. 246. In Suter's transcription the reference to *nullus* in connection with *cifra* is replaced by *nihilum* in Manuscript O. See n. 10, section I, p. 10. For number forms see Rita Irani, 'Arabic Numeral Forms', *Centaurus* 4 (1955), pp. 1–12; A. Allard, 'L'époque d'Adélard et les chiffres arabes dans les manuscrits latins d'arithmétique', *WIST* XIV, pp. 37–43, figs I and II, pp. 40, 41; R. Lemay, 'The Hispanic Origin of our Present Numeral Forms',

Viator 8 (1977), pp. 435–67. Lemay draws attention to the use of 't' for *teca* (*theta*, θ) or Latin *terminus*, and τ as *cifra*, the symbol for zero. See also Richard Lemay, 'Arabic Numerals' and 'Roman Numerals', *DMA* vol. 1, pp. 382–98, vol. 10, pp. 470–74.

19 Measurement of degrees of latitude in camel paces and miles: Neugebauer 1962, I, pp. 7, 8, 9; II, pp. 17, 18, 19.

20 Petrus Alfonsi, *Dialogi cum Judaeo*, *WIST* XIV, Catalogue 13, p.169; J.P Migne, *Patrologia Latina*, vol. 157, cols 535–672. Oxford, Corpus Christi MS 283, ff. 113–44; Neugebauer, 1962, Appendix, pp. 216–18; E.J. Kealey, *Medieval Medicus* (London and Baltimore, 1981), p. 197, n. I am indebted to Professor Kealey for permitting me to read his draft study of Adelard, Petrus and Walcher of Malvern in which he considers a possible association. See also Charles Burnett, 'Traductions et Traducteurs de l'antiquité tardive au XIVe siècle', *Rencontres de cultures dans la philosophie médiévale*, Publications de l'Institut d'Études Médiévales (Louvain-la-neuve, Cassino, 1990). Burnett suggests that Petrus was the teacher from whom Adelard learned Arabic.

21 Mercier, pp. 99, 100. Mercier states that his conclusions may be seen as a development of arguments put forward by J.M. Millás-Vallicrosa, in 'La aportación astronómica de Pedro Alfonso', *Seferad* 3 (1943), pp. 65–105. See also Dorothea Metlitzki, *The Matter of Araby in Medieval England* (New Haven, Conn., 1977), pp. 25,26,27.

22 Neugebauer 1962, II, p. 234.

23 Orbits of planets, Hindu tradition: Neugebauer 1962, II, p. 14.

24 Longitude of planets according to Hindu method: Neugebauer 1962, II, p. 23.

25 Lunar motion not according to *Almagest*: Neugebauer 1962, I, p. 10; II, p. 21.

26 Importance of understanding variations: Neugebauer 1962, II, p. 14. Richard Lemay has informed me of Hermann of Carinthia's comment on tables in his translation of *Almagest*. Lemay emphasises that Hermann was the translator of *Almagest* from the Greek, a translation once attributed tentatively to Adelard of Bath. Franz J. P. Bliemetzrieder, *Adelhard von Bath, Blätter aus dem Leben eines englischen Naturphilosophen des 12. Jahrhunderts und Bahnbreckers einer Wiedererweckung der Griechischen Antike* (Munich, 1935), pp.149–274; Haskins, *Studies*, 1927, pp. 158, 159; Richard Lemay, 'De la scolastique à l'histoire par le truchement de la philologie: itinéraire d'un médiéviste entre Europe et Islam', estratto dagli atti del Convegno Internazionale promosso dall'Accademia Nazionale dei Lincei-Fondazione Leone Caetani e dall'Universita di Roma 'La Sapienza' Facolta di Lettere – Dipartimento di Studi Orientali, *La difffuzione delle scienze islamiche nel medio evo europeo* (Roma, 2–4 ottobre 1984). Roma (Accademia Nazionale dei Lincei, 1987), pp. 399–533 (429–62).

27 Reference by Adelard to *Almagest* in connection with sines: Neugebauer 1962, I, p. 18; II, p. 45.

28 Mercier, p. 98. I am indebted to Professor Mercier for elaborating on this point for me.

29 I am indebted to the late Professor Eric Forbes of Edinburgh University for drawing the al-Bīrūnī experiment to my attention. See N.A. Baloch, *Beruni and his Experiment at Nandana* (Islamabad, 1983).

30 Suter 1914, p.99. J.K. Wright (n. 14 above) pp. 83,98 states that men of the Latin West wanted to know the astronomical positions of cities in order to transpose solar, lunar and other tables based on the meridian of one city to the meridian of others. Accuracy in geography was of no particular concern. The main interest to the medieval mind of this data was astrological.

31 *Liber ysagogarum Alchorismi*, described in *WIST* XIV, Catalogue 23, pp. 173, 174. Eight manuscripts are listed. The work is described as 'in five books attributed in P, *Paris, BN, lat.* 16208, (interpolated version) to 'Magister A', anonymous in other manuscripts. The work is based on Boethius *De arithmetica*; a translation of al-Khwārizmī on Indian calculation; Christian and Jewish computistical works; a *Theorica planetarum*; a translation of Euclid's *Elements* covering parts of the work surviving in the Boethian tradition together with items not found hitherto in that tradition .... The appearance of a Jewish calendar and the equivalents of the nine digits among the letters of the Jewish alphabet, the Hebrew names of the planets, and a table of eras which is almost identical with that in the calendrical tables following Petrus Alfonsi's *Letter to the Peripatetics of France* suggest that Petrus was the author of at least that book.'

The first three books, on the algorism of

al-Khwārizmī, ed. M. Curtze, 'Über eine Algorismus-Schrift des XII. Jahrhunderts', *Abhandlungen zur Geschichte der Mathematik* 8 (1898), pp. 1–27; see also A. Allard, *Les Plus Anciennes Versions latines du xiiè siècle issues de l'Arithmétique d'Al-Khwārizmī*, Centre d'ecdotique et d'histoire des sciences, Université de Louvain (Louvain, 1975), pp. 17–19. See also n. 34 below.

32 Charles Burnett, 'Introduction', *WIST* XIV, pp. 4, 5, concludes that given the eclectic nature of the text, it is more likely that the author is an unknown compiler who has used texts associated with Petrus Alfonsi alongside other texts. Folkerts, *WIST* XIV, p. 63, expresses the view that Adelard is 'Magister A'. Allard, *WIST* XIV, p. 38, states that the safe view is that 'Magister A' is neither Petrus Alfonsi nor Adelard of Bath. Richard Lemay had at one time thought 'Magister A' was Petrus Alfonsi, more recently that he could have been Abraham ibn Ezra or Abraham Savasorda, (letter to Louise Cochrane, 17 June 1984).

33 *De numero Indorum*, *WIST* XIV, Catalogue 9, p. 168, Cambridge University Library MS Ii.6.5, ed. A.P. Juschkewitsch, 'Über ein Werk des Abdallah Muhammed ibn Musa al Huwarazmi al-Magusi zur Arithmetick der Inder', *Schriftenreihe für Geschichte der Naturwissenschaften, Technik und Medezin, Beiträge zur Geschichte der Naturwisssenschaften, Technik und Medizin herausgegeben zum 60. Geburtstag Gerhard Harigs*, ed. I. Strube and H. Wussing (Leipzig, 1964), pp. 21–63 (facsimile of MS on pp. 22,23 and 55–63); ed. K. Vogel, *Mohammed ibn Musa Alchwarizmi's Algorismus des fruheste Lehrbuch zum Rechnen mit Indischen Ziffern* (Aalen, 1963); A. Nagl, 'Über eine Algorismusschrift des XII. Jahrhunderts und über die Verbreitung der indisch-arabischen Rechenkunst und Zahlreichen in christlichen Abendlande', *Zeitschrift zur Mathematik und Physik* 34 (1889), Supplement, pp. 129–46, 161–70; M. Curtze, n. 31, above.

34 See ch. 3, above. Richard Lemay, 'Arabic Numerals' in *DMA*, vol. 1 (1982), pp. 382–98; 'Roman Numerals', *DMA*, vol. 10 (1988), pp. 470–4; Michael Mahoney, 'Mathematics', *DMA*, vol. 8 (1987), pp. 205–22; Guy Beaujouan, 'Medieval Science in the Christian West', in René Taton (ed.), *Ancient and Medieval Science, A General History of the Sciences*, 4 vols, (London, 1963), vol. 1, p. 480. Lemay 1977, 'Numeral Forms', n. 18, above; S. Gandz, 'The Origin of the Ghubār Numerals or the Arabian Abacus and the Articuli', *Isis* 16 (1931), pp. 393–424; Gillian Evans, 'From Abacus to Algorism', *British Journal of the History of Science* 16 (1977), pp. 114–31; Harriet Lattin, 'The Origin of our Present System of Notation according to the theories of Nicholas Bubnov', *Isis* 19 (1933), pp. 181–94.

35 A broadsheet from *FWC* (1980), gives examples of thirteenth-century arabic numerals used by masons on sculptured figures.

CHAPTER 9

Adelard and Astrology

The sudden upsurge in interest in astrology in the twelfth century was the result of the translations from the Arabic which originated in Spain. Adelard had expressed his own interest in astrology early in his career.[1] He was already sufficiently knowledgeable to seek out the specific information he required during his journey to Syria or acquire information from other sources. Among the Arabs astrology was considered to be a science closely related to astronomy. The texts Adelard selected for translation were *Centiloquium Ptolemei*, 100 aphorisms to summarise the fruits of Ptolemy's astrology; the *Isagoge minor* ('Shorter Introduction to Astronomy' that is, Astrology) of Abū Ma'shar; and the *Liber prestigiorum Thebidis*, a book on the theory of images, by Thābit b. Qurra. It is possible that Adelard also translated a short text on chiromancy from the Greek.[2]

Richard Lemay suggests that Adelard probably began his work of translation from the Arabic not with Euclid's *Elements* but with the *Centiloquium*:

> This translation is very deficient for lack of understanding of both the Arabic and the 'science' of astronomy in Adelard when he undertook it. He left it unfinished before he reached midway through the 100 propositions (he ended with prop. 39, in fact) and he never undertook the commentary which is the essential and most substantial part of the *Centiloquium* composed by Ahmed ibn Yusuf about the year 920 AD. Adelard's attempt seems to me, therefore, to be his first encounter with an Arabic text and must be placed early in his 'Arabic' career. The *Ysagoge minor* which is also somewhat deficient with respect to the original Arabic though less so than the *Centiloquium* must come next as well as the translation of Thābit's *Prestigia*. The tables of al-Khwārizmī (probably 1126) followed, and then probably the Euclid...[3]

It seems quite reasonable to consider that Adelard attempted translations during the period of his travels. After all soon after his return he wrote *Quaestiones naturales* in which he describes himself as having a ring set with an emerald, possibly the sort of talisman described by Thābit b. Qurra. Incidentally in the *Liber prestigiorum* Adelard makes a reference to Bath, using it instead of Baghdad as the example of a city from which scorpions are to be driven out.[4]

The question that is difficult to answer is how extensive was Adelard's knowledge of Arabic? Was his understanding predominantly aural and did he have an Arabic scholar to assist him as he worked on the texts? This could have been the case during the time he was abroad. It is felt that by the time Adelard worked on translations of Euclid's *Elements* and al-Khwārizmī's *Zij* he had returned to England. He probably became associated with the circle of West Country mathematicians who followed in the wake of Robert, Bishop of Hereford, and his successor, Gerard, who became Archbishop of York. Both were extremely interested in astrology. The group had also included Walcher, Prior of Malvern, who had worked with Petrus Alfonsi. A high proportion of the scholars who went to Spain in search of Arabic science were English; help with Arabic must have been available. Adelard's understanding of mathematics and astronomy was certainly as important as his understanding of Arabic.[5]

What was the subject-matter of the earlier translations? Adelard would have assumed that the original purpose of the *Centiloquium* had been to lay down the general principles in Ptolemy's *Tetrabiblos* ('Four Books') as a guide to Ptolemy's astrological theories. Ptolemy had approached astrology from the point of view of an astronomer. His astrology was compiled during the last stages of Greek astrology. He did not use Aristotle's theory of motion to justify his interpretation, but he did borrow from *De generatione et corruptione*.[6]

At the outset in *Tetrabiblos* Ptolemy made clear that in his view the perceptions of astrology should not be compared with the unvarying science he had presented in the *Almagest*, and that in this work he would point out the weakness and unpredictability of material qualities found in individual things as compared with the certainties of the movement of the stars. According to Ptolemy, a certain power emanates from the aether, causing changes in the sublunar elements, and in plants and animals. John North explains this:

> Effluence from the Sun and Moon – especially important was the Moon, by virtue of her proximity – affects things, animate and inanimate, while the planets and stars have their own effects. Given the key to calculating these effects – and that is what the *Tetrabiblos* is all about – one might be able to work out the weather and human character in advance.[7]

Ptolemy divided his work into four parts, the first outlining the principles, the second dealing with general predictions regarding war, pestilence, earthquakes, floods, storms, hot and cold weather and fertility. The last two books dealt with predictions regarding individuals.

For the purpose of general predictions Ptolemy divided the inhabited parts of the earth into seven 'climates' (related to degrees of latitude), each of which was governed by particular constellations and planets. The earth's surface was also subdivided into quadrants subject to various influences:

> Now of the four triangular formations recognised in the zodiac, as we have shown above, the one which consists of Aries, Leo, and Sagittarius is north-western and is chiefly dominated by Jupiter on account of the north wind, but Mars joins in its governance because of the south-west wind ... Under this arrangement, the remainder of the first quarter, by which I mean the European

> quarter, situated on the north-west of the inhabited world, is in familiarity with the north-western triangle, Aries, Leo and Sagittarius, and is governed as one would expect by the Lords of the triangle, Jupiter and Mars, occidental [in a western position]. In terms of whole nations these parts consist of Britain, [Transalpine] Gaul, Germany, Bastarnia [the south-western part of Russia and southern Poland], Italy [Cisalpine Gaul], Apulia, Sicily, Tyrrhania, Celtica and Spain. As one might expect it is the general characteristic of these nations, by reason of the predominance of the triangle and the stars which join in its government, to be independent, liberty-loving, fond of arms, industrious, very war-like, with qualities of leadership, cleanly and magnanimous.[8]

In Ptolemy's general predictions the advent of an eclipse or a comet could be said to forebode evil for this or that country according to the constellation in which it first appeared. Ptolemy felt this type of prediction was the surest part of his science. In practice it was less important than prophecy regarding the individual, related to the date and time of his birth: his parents, brothers, length of life, health, riches, profession, marriage, children and friends. This is judicial astrology.[9]

The *Tetrabiblos* offers a complete summary of astrological ideas as they were understood at the time Ptolemy was writing. It is a well structured and detailed work. The *Centiloquium*, by contrast, is intended to represent the fruits of Ptolemy's astronomy in 100 propositions. For example:

> IX. In their generation and corruption forms are influenced by the celestial forms, of which the framers of talismans consequently avail themselves, by observing the ingresses of stars thereupon.
>
> XIX. The efficacy of purgation is impeded by the Moon's conjunction with Jupiter.
>
> XXII. Neither put on nor lay aside any garment for the first time when the Moon may be located in Leo. And it will be still worse to do so, should she be badly affected.

Adelard probably believed that he was translating a summary which Ptolemy had written based on his own work, but this was not the case. Lemay has shown that the *Centiloquium* was in fact the work of Ahmed ibn Yusuf, an astrologer from Cairo, *c.* 920. It was a forgery, entitled *Kitāb Thamara.*[10]

Adelard turned his attention next to the work of Abū Ma'shar, and translated the *Isagoge minor* ('Shorter Introduction to Astronomy'). This was a better choice than the *Centiloquium* in that Abū Ma'shar had done his own summary of the *Introductorium maius* ('Greater Introduction to Astronomy'). Adelard's translation of the briefer work thus provided the initial impetus for Abū Ma'shar's tremendous influence in the Latin West.

Abū Ma'shar's name became so well known over the centuries that it was used for the character of the cheating astrologer in a play performed for King James I (and VI of Scotland) at Trinity College, Cambridge, in 1614. At that time certain aspects of astrology had fallen from royal favour. But a true representation of Abū Ma'shar must credit him with being one of the foremost Arabic astronomer-astrologers. He wrote both the *Ysagoge minor* and the *Introductorium maius* during the period when Aristotle's works were being translated into Arabic at the House of Wisdom. As a result many Aristotelian ideas found their way into his texts, and thence, in Adelard's time, to Europe.[11]

Aristotle had been responsible for revising Plato's theory of the twofold influence on sublunary matter. The perfect diurnal motion of fixed stars from east to west constitutes the principle of permanence and growth, whereas the motion of the planets running their annual courses at irregular paces from west to east athwart the diurnal motion constitutes the principle of earthly change. When once interpreted astrologically Aristotle's physics of motion was all that was needed to fasten upon the Middle Ages the exaggerated belief in the importance of the stars which lay at the basis of faith in astrology.[12]

Abū Ma'shar had borrowed from *De caelo*, *De generatione et corruptione*, *Meteorologica*, *Metaphysica* and *Physica*. There is no attribution but the ideas are inherent in the work. But Abū Ma'shar stretched Aristotle's interpretation to suit his astrology. For example, Aristotle in looking for an explanation of the diversity of influence related to the planetary system attributes variety to the effect of the ecliptic. Abū Ma'shar extends this to account for astrological lore.[13]

Abū Ma'shar's complete astrology was contained in the work he called the *Introductorium maius*, translated not by Adelard but by John of Seville in 1133 and Hermann of Carinthia in 1140.[14] Hermann's *De essentiis* was truly impregnated with ideas of Aristotle's natural philosophy, mostly derived from Abū Ma'shar. His work emphasises the difference medieval thinkers believed to exist between the sublunar world composed of the four elements and the upper heavens where celestial bodies were made of unchanging essence which could receive no alteration. How then do these heavenly bodies influence the sublunar world? There are five genera of the *Essentia* – *Causa*, *Motus*, *Locus*, *Tempus* and *Habitudo*. Their role is to bring all other things into existence. They have in themselves the nature of the same and yet are the roots of diversity. Here Hermann is trying to merge the Boethian notion of *Essentia* as an unalterable form with the Platonic dichotomy of the Same and the Different. Hermann's substance is the *Diversum*.

Heavenly bodies formed of *Essentia* (representing the Same) differ from one another in terms of the five genera. *Causa* has an original creative meaning of a theological nature. What the theory amounts to is that God, having delegated creative capacity to the heavenly bodies (*Causa*), also endows them with permanently unchanging characteristics under the above headings, and they in their activity influence events of the sublunar world and are responsible for all coming into being and passing away. In other words heavenly bodies have a simple and unchanging nature which never alters; they are forever in motion, but their varieties are innumerable and can be reduced to five genera. *Habitudines* are the conditions included in the list of variations which can be applied to heavenly bodies; it is because of the variety of their influence on events on earth that they can be said to account for all generation and passing away.[15] There were twenty-five different conditions or *Habitudines* which could be allocated to different planets – these presumably provided the traditional characteristics associated with each one.

It is easy to see how theories of this nature could lead to fatalism and to a pantheistic cosmology – a highly dangerous doctrine from the point of view of the Church, especially in the implication that the soul has no control over its own destiny. So although some of Aristotle's metaphysical ideas found their way to the West through astrology, at a later date it was necessary to dissociate Aristotelianism from some of the astrological interpretations.[16]

Adelard did not give serious attention to Abū Ma'shar's full text. He ignored, for example, the explanation of tides as related to the moon. What he stressed was the close relationship between astronomy and astrology (Lemay's translation):

> Whoever wishes to make a constant study of the higher scientific philosophy seeking at the same time a sensible explanation of the celestial universe and wishes to discover the effect on the aetherial world must emphatically take note of the portents of the circles of the signs. He must be aware of which signs might be master in particular points, and he must be able to foretell which are chemelia [northern] and which genubia [southern] . . . .[17]

The *Isagoge minor* which Adelard translated contains in effect only the rules which govern horoscopes. It lists the signs of the zodiac and their order, gives the names of the planets and describes the various houses. It gives the qualities of the various signs – hot, dry, moist, earthy – and their influence over things in the lower world. It includes a description of men's limbs in relation to the constellations, a highly popular conception of man as microcosm.

Adelard's translation was not as important to European scholars as was the full text presented by John of Seville (who called Adelard the 'fatuous Antiochean') and later that of Hermann of Carinthia. The *Introductorium maius* was used by Gerard of Cremona in public lectures on astrology in Toledo. Hermann's version later became the source of the first printed text.[18]

Hermann of Carinthia was a pupil of Thierry of Chartres and was acquainted with Adelard's translation. He also edited a text of the *Elements* as well as *Sphaerica* of Theodosius from what might have been another text of Adelard's. There appears to have been some relationship between the works of these two men. A version of *De essentiis* by Hermann was thought at one time to be the explanation of God promised by Adelard at the end of *Quaestiones*. Hermann was anxious to produce a synthesis of Arabic ideas which by the middle of the century were well known in Europe. He drew up a list of scientific books.[19]

Adelard would not consider astrological interpretations to be less scientific than his other pursuits. No one can understand even the simplest claims of astrology without some grasp of astronomical principles. In the casting of horoscopes the first essential is to know the position of sun, moon, planets and stars for the exact moment of a person's conception or birth, or for the founding of a city. The year is divided into twelve signs – Aries, Taurus, Gemini and so on – in accordance with which is on the meridian at noon. Each sign is subdivided into 30° so lasts approximately a month. During every twenty-four-hour period the position of the stars overhead is constantly changing. Some signs are in the ascendent, rising in the east, some are descending, some are in opposition. Everyone is born into one of twelve 'houses'. Each of these is designated by a pre-determined section of a diagrammatic map of the heavens. For every individual the choice of his 'house' is established in accordance with the time of day as well as the date of his birth. Planets too have houses and are in different signs at different times. This is why one can say there are innumerable possibilities.

The Arabic terms *genubia* and *chemelia*, which Adelard leaves unexplained in his translation of Abū Ma'shar, receive detailed explanation in the *Zij* and again in the treatise on the astrolabe. They are straightforward Arabic terms for south (*janūb*) and north (*shamāl*) and refer to the altitude or declination of the sun at noon on the meridian in the period of the sign. The sun is north of the celestial equator for Aries, Taurus, Gemini, Cancer, Leo and Virgo, so northern signs are *chemelia.* Those called *genubia* are southern: Libra, Scorpio, Sagittarius, Capricorn, Aquarius, Pisces. One must constantly remind oneself that to appear south of the celestial equator means the sun is lower in the sky from the point of view of observers in our latitude. It is north of the equator during spring and summer. Adelard explains in Chapter 15 of the *Zij*, 'On the Knowledge of the Declination of the Sun':

> What is found between the first degree of Aries and the third sign is called *shemeli*, that is on this northerly side, and increasing: if however between the third sign and the sixth, *shemeli* and decreasing. If between the sixth and the ninth it is called *genubi* (that is on the other side southerly) and decreasing – if it is between the ninth and the twelfth sign it is called the right side and increasing.

Adelard makes sure that those using the *Zij* understand what is involved. At the time of the equinox, the first degree of Aries or the first degree of Libra, when the ecliptic intersects the equator, we see the sun in the sky in accordance with the latitude from which we are making the observation. If we are in latitude 52°N., the sun will be 38° above the horizon. But because of the obliquity of the ecliptic, we have to do a more complicated sum at all other times. During *chemelia* the obliquity is added to the declination, during *genubia* it is subtracted.[20]

Al-Khwārizmī's *Zij* is replaced in modern times with an ephemeris for astrology as well as astronomy or navigation. With this as handbook the expert can establish where various heavenly bodies are at any given time, past, present or future, providing he has an ephemeris for the correct year. But the vocabulary and sign language of astrology are such that those who are interested have to make a special study of what is involved. Adelard's translations of astrological texts not only introduced ideas but helped to perpetuate many beliefs which are quite familiar to those who like to know what the stars foretell in their lives, whether they put credence in it or not.

Modern astrology carries on many of the early techniques. The diagrammatic map of the heavens is constructed, often based on a square. This is in fact merely a device for recording the information the astrologer must collate. There are more than a dozen ways of dividing a horoscope, but there is one method of supreme importance, a method that was known throughout the Middle Ages, and is considered 'standard'.[21] It is the one used by Adelard and will be discussed in Chapter 10, in connection with the astrolabe treatise.

Having recorded where the signs of the zodiac are in relation to the horoscope, one must work out the exact position of sun, moon and planets for the time and date in question. The astrologer is then able to analyse the many interacting influences which affect the horoscope. These include 'aspects': trine for signs 120°, sextile for 60°, quintile for 72° and quartile for 90°. Trine and sextile are always good; quintile is slightly good; square or quartile is always evil. Conjunction means two signs appear in the same house. Opposition means they are 180° apart.[22]

Obviously there were concerns about various influences and whether anything could be done to encourage good ones and counteract those which seemed unfavourable. This too was a subject on which Adelard found a text to translate, the *Liber prestigiorum Thebedis secundum Hermetum et Ptolomeum* ('The book of wonderful works of Thābit following [the authority of] Hermes and Ptolemy'), of Thābit b. Qurra.

By Hermes he means Hermes Trismegistus whose astrology was neo-Platonic. Authorship of the so-called Hermetic books was attributed to the Egyptian god of wisdom, Thoth, whose name was sometimes translated into Greek as Hermes Trismegistus ('Thoth the thrice great') and was therefore equated with the Greek god Hermes. The legend was that he was a priest-lawgiver of the ancient Egyptians who had been the source of Plato's Gnostic ideas. It was later proved that the corpus was written by various authors between the second and fourth centuries and contained Jewish, Egyptian and Persian ideas.

The Hermetic corpus consisted of fifteen books concerned with the occult and with the practice of talismanic magic for drawing down influence from the stars and making contact with daemons and angelic powers outside the earth. They included an elaborate description of the sympathetic virtues of certain flowers, plants and stones. Relics of Hermeticism persist in astrological practices today, for example, in the belief in lucky birthstones. Adelard's ring mentioned in *Quaestiones* could have had this sort of significance.

The theory was that the influence of the stars could be drawn down by feats of sympathetic magic (*praestigia*) accomplished through astrological practices. In the Middle Ages practically every object of the material world was considered to have occult sympathies with the star on which it depended. In order to capture the power of Venus one had to know which stones or metals or animals or plants were related to Venus, how to make an image of Venus, and when was the right moment to do it.

All constellations and planets were controlled by angelic powers, and astrological knowledge was used to reinforce or to contradict the indications which astrologers produced by casting horoscopes. The composite authors of the *Corpus Hermeticum* were strongly influenced by Greek thought and were post-Christian. There was a danger, however, in the fact that the doctrines were mixed with an Egyptian and Persian element which led directly to astrological daemonic magic. Christian thought is not evident and Augustine condemned Hermes, the Egyptian, called Trismegistus, for the idolatry and magic found in some of his writings.[23]

The work of Thābit b. Qurra which Adelard translated was a summary of the theory of images. This work was an unimportant one compared with Thābit's translations of mathematical texts. It describes the making of seals or amulets which are astronomical and astrological and must be constructed under prescribed constellations to achieve the ends sought. Abū Ma'shar, who was a near contemporary of Thābit during the ninth century, recorded that in Baghdad at this time the making of amulets was a thriving industry. Thābit emphasises the need for a knowledge of astronomy to perform feats of magic. The images are often human forms rather than astrological figures. It is not necessary to engrave them on gems. Thābit expressly states that the material on which they are engraved is not important and that lead, tin, bronze, gold, silver, wax, mud or anything will do. The essential thing is the perfection achieved in careful conformity to astrological conditions, hence the predominant role of 'astrological force' rather than the material used. Yet in the commentary to Proposition IX of the *Centiloquium*, Ahmed ibn Yusuf tells the story of a Greek physician having settled in Cairo who was an expert on the virtue of stones. He cured his squire's scorpion bite by crushing a powder with a stone amulet engraved with the image of a scorpion, and the squire who was made to 'drink' the product fell asleep and when awakening later was totally cured. One suspects some anaesthetic substance to have been involved, and hence the material would have been important.[24]

The translation of a short work on chiromancy (palmistry) is also sometimes attributed to Adelard. The text appears in the same manuscript at the British Library as do the *Centiloquium* and the *Isagoge minor* of Abū Ma'shar. It is illustrated by two extraordinary diagrammatic hands with instructions on how to 'read' them written across the palms and along the fingers. The text has no Arabisms and was translated from the Greek. The author named is Aristotle – elsewhere Adelard's name appears as translator.[25]

These works are not comparable with Adelard's translation of Euclid or of al-Khwārizmī's *Zij*. Lemay considers that their impact was negligible. Burnett suggests that Adelard is grappling with the problem of how to introduce new sciences to a Latin public and this might account for some of the stilted language. Lemay has also pointed out that from Adelard's time onwards astrology began to be treated as a superior branch of physics, a sort of provisional metaphysics, to be later displaced in the thirteenth century.

Recently Burnett has drawn attention to a manuscript entitled *Ut testatur Ergaphalau*, a work on the position of the science of the stars within a general division of knowledge. It appears together with Adelard texts. The author of this work believed that *Megacosmica* or *Astronoida*, the opposite of *Microcosmica* or *Medicina*, was a subdivision of *Phisica* (in its turn a subdivision of *Sapientia*). *Megacosmica* had two branches, *Astronomia* and *Astrologica*. These in their turn accounted for further subdivisions.

Burnett considers that *Ut testatur Ergaphalau* ('The Science of the Stars') may have been deliberately associated by Adelard with his own works on astrology. The unknown author may have had some connection with the astrologer whose book, known as the *Liber Alchandrei*, was said by William of Malmesbury to have inspired Gerbert's pact with the devil. Burnett believes that a study of both *Ut testatur Ergaphalau* and *Liber Alchandrei*, together with Adelard's early astrological works, could show 'one way in which the traditional division of the quadrivium was set aside and replaced by a division of sciences more accommodating to astrology'.[26]

It is interesting, moreover, that Adelard's choice of subject-matter and of authorities to translate seemed very relevant to succeeding generations, even though his early translations were less successful than those of other Arabists. Astrology was a recognised subject for study, although chiromancy, palmistry, magic and geomancy were not. Abū Ma'shar's *Introductorium maius* and the spurious *Centiloquium* continued as texts in the quadrivium. J. V. Field reports that at the University of Krakow in 1525 the astrological works being studied in an astronomy course included Ptolemy's *Tetrabiblos* and *Centiloquium* as well as treatises by Abū Ma'shar and other astrologers. There was also a revival of interest in Hermeticism during the Renaissance.[27]

A new dimension has been added recently to the portrait of Adelard which historians have gradually been assembling. It is the result of the attribution to Adelard, by John D. North, of ten royal horoscopes.[28] Previously Adelard's association with the court has been thought of as a modest one perhaps connected with the Exchequer. The possibility that he acted as astrologer is of great interest. Certainly the study of his works has led increasingly to the realisation that Adelard viewed astrology as equal in importance to astronomy. It would have been most surprising if he did not avail himself of the tables he had translated and the techniques he had acquired. The dating of most of these horoscopes as having been cast in the last years of Stephen's reign and the fact that al-Khwārizmī's *Zij* was used support the suggestion that Adelard was the astrologer. It is, however, very curious that no attempt was made to allow for a difference in latitude or longitude between Cordova and England.

If Adelard is the astrologer, it means that he lived on into the second half of the twelfth century. The horoscopes probably survived because of their royal associations. To be able to link them with known events in Stephen's reign has helped to confirm their authenticity. Adelard's advanced age and his reputation would render his conclusions authoritative. The horoscopes are of the type described as 'elective' – the way of discovering the propitious moment for a certain action – or 'interrogative' – in order to discover the answer to a question. The translations from the Arabic which had occurred in the twelfth century had introduced a new element into astrology. Ptolemy had been concerned almost entirely with judicial astrology; he had ignored these two other aspects of great interest to the Arabs – interrogation and election.[29]

What are the questions which are posed? In 1151 the King was advised to persuade his barons to swear fealty to his son. Stephen's son Eustace was to have succeeded him but there was strong feeling that Matilda's son, Henry Plantagenet, was the rightful heir to the Crown. A propitious moment was required for swearing such an oath. A horoscope was cast to decide this.

Would the Normans (the Plantagenets) come to England? This was another crucial question on several occasions when there was a threat of invasion in support of Henry's cause. Geoffrey Plantagenet did not come but Henry himself did. After Eustace's death in 1153 Stephen disinherited his second son William and made Henry his heir in the Treaty of Winchester.

One horoscope records a meeting between a master and former pupil – Adelard and Henry II in all probability. Adelard's astrolabe treatise was almost certainly dedicated to the young prince either in 1142/3 or nearer to the time of Henry's coming of age in 1149.[30]

Is it possible to associate Adelard with both sides of the civil war between Stephen and Matilda in this fashion? The answer can be yes because Bath was virtually on the boundary between the strongholds of two opposing factions. The Bishop of Bath at this time was Robert, a protégé of Stephen's brother, Henry of Blois, Bishop of Winchester and Abbot of Glastonbury. Bishop Robert was probably the anonymous author of *Gesta Stephani*. He describes in some detail the threat to the bishop's life (his own) when Robert of Gloucester's forces convinced him of the consequences if there were no surrender.[31] Later he reports the meeting between Stephen and the bishop when the city is retaken. It is suggested that life in Bath was probably similar to that under enemy occupation since there were so many resident supporters of Robert of Gloucester, leader of his half-sister Matilda's forces. This would justify the claim that Adelard was summoned as tutor in mathematics for Henry during his stay with his uncle at Bristol, especially since Adelard must have been a man of renown among his contemporaries.

By the time of the horoscope concerning a master and a former pupil, Stephen's position was far from secure and Adelard would have been more than willing to see the prince again. He would not have been the only one to wish to be considered favourably by the possible heir to the throne.

Of particular interest also, in view of the nature of the horoscope manuscripts, is the hypothesis put forward by John North that the document containing these horoscopes must be an Adelard autograph.[32] All but one are written in the same hand.

Notes

1 *De eodem*, pp. 31, 32. See above, ch. 2, pp. 17, 19 and n. 16, p. 20.

2 Adelard of Bath, *Centiloquium Ptolemei*, *WIST* XIV, Catalogue 1, p. 166; *Isagoge minor* of Abū Ma'shar, *WIST* XIV, Catalogue 18, p. 172; *Liber prestigiorum Thebidis*, *WIST* XIV, Catalogue 21, p. 173; Charles Burnett, 'Adelard, Ergaphalau and the Science of the Stars', *WIST* XIV, pp. 133–6, discusses these three texts, pp.133–136; *Ciromantia*, *WIST* XIV, Catalogue 2, p. 166.

3 Letter from Professor Richard Lemay to Louise Cochrane, 17 June 1984. Professor Lemay was kind enough to reply to a query from me about the chronological order of Adelard's work.

4 *Quaestiones*: see ch. V, n. 39. Gibson, *WIST* XIV, p. 16. Charles Burnett, *WIST* XIV, p. 135, n. 10, citing MS Lyon, Bibliothèque municipale, 328, f. 73v., Catalogue 70, p. 184.

5 For existence of a circle of West Country mathematicians interested in astrology see Sir Richard Southern, *Medieval Humanism and Other Studies* (Oxford, 1970), pp. 169–71; Gibson, *WIST* XIV, p. 15; Brian Lawn, *The Prose Salernitan Questions* (London, 1979), Introduction, pp. xv-xvii.

6 Theodore Wedel, *The Mediaeval Attitude towards Astrology particularly in England*, Yale Studies in English 60 (New Haven, Conn., 1920), pp. 2, 3.

7 Claudius Ptolemy, *Tetrabiblos*, ed. and trans. into English F.E. Robbins, Loeb Classical Library (London and Cambridge, Mass., 1940) Book I, Introduction, p. 3; John D. North, 'Medieval Concepts of Celestial Influence: A Survey', in Patrick Curry (ed.), *Astrology, Science and Society* (Woodbridge, Suffolk and Wolfeboro, New Hampshire, 1987), pp. 5–17 (6).

8 *Tetrabiblos*, Robbins, Book II, 1, p. 121, 'climates'; 3, pp. 129–36, 'triangular formations' cross-referring to Book I, 18, p. 83 .

9 Wedel, pp. 8–10.

10 For English translation of *Centiloquium* quoted here see J.M. Ashmand, *Ptolemy's Tetrabiblos or Four Books on the Influence of the Stars (Contains Extracts from the* Almagest *and the Whole of his Centiloquy – One Hundred Aphorisms of Claudius Ptolemy, Otherwise called the Fruit of his Four Books*) (London, 1917), pp. 225, 226. For authorship of *Centiloquium* see Richard Lemay, 'Origin and Success of the *Kitāb Thamara* of Abū Ja'far Ahmad ibn Yūsuf ibn Ibrāhīm', in *Proceedings of the First International Symposium for the History of Arabic Science, April 5–12, 1976*, 2 vols (Aleppo, 1978), vol. 2, pp. 91–107 (101).

11 Richard Lemay, *Abū Ma'shar and Latin Aristotelianism in the Twelfth Century* (Beirut, 1962), *passim*, Introduction, p.xxiv. For the play see Thomas Tomkis, *Albumazar, A Comedy*, ed. Hugh Dick (Berkeley, Calif., 1944), first produced on the stage of Trinity College, Cambridge, 9 March 1614/15 on the occasion of a visit by King James to the University. It represents the anti-astrological literature of the Renaissance. Richard Lemay comments that it 'plagiarizes' G. della Porta, *L'astrologo*. Note accompanying letter to L. Cochrane, 11 December 1990.

12 North 1987, pp. 5–16; Wedel, p. 5.

13 Lemay 1962, p. 83.

14 Lynn Thorndike, *A History of Magic and Experimental Science* 2 vols (New York, 1923), vol. 1, p. 652; Charles Burnett, 'Arabic into Latin in Twelfth-Century Spain: The Works of Hermann of Carinthia', *Mittellateinisches Jahrbuch* 13 (1978), pp. 100–34; David Pingree, 'Abū Ma'shar Al-Balkht Ja'far ibn Muhammad', *DSB*, vol. 1 (1970), pp. 35–9.

15 Lemay 1962, pp. 6, 199, 200. Hermann of Carinthia, *De essentiis*, ed. Charles Burnett (Leiden, 1982).

16 Wedel, p. 16.

17 Lemay 1962, p. 7, translation of Adelard in Appendix.

18 Lemay 1962, pp. 3–9.

19 Burnett 1978, pp. 104–6.

20 Neugebauer 1962, II, p. 33. Neugebauer adds a note to explain that 'Right side' refers to the right side of the equator in the direction of the daily rotation.

21 John D. North, *Horoscopes and History*, *WIST* XIII (1986); 'Some Norman Horoscopes', *WIST* XIV, pp. 147–61 (147); 'Celestial Influences', ch.18, *Stars, Minds and Fate, Essays in Ancient and Medieval Cosmology* (London, 1989); Emmanuel Poulle, 'Le traité de l'astrolabe d'Adélard de Bath', *WIST* XIV, pp. 128, 129.

22 *Tetrabiblos*, Robbins, Book I, 13, pp. 73, 75; Wedel, p. 16.

23 See ch. 4, pp. 36–7, for connection between *Mappae clavicula* and Hermes Trismegistus. See also Karl H. Dannenfeld, 'Hermes Trismegistus', *DSB*, vol. 6 (1973), pp.305, 306.

24 Thorndike 1923, pp. 664, 665; B.A. Rosenfeld, A.T. Grigorian, 'Thābit b. Qurra', *DSB*, vol. 13 (1976), pp.288–95; quotation from Yusuf contributed by Lemay, note with letter to L. Cochrane, 11 December 1990.

25 Lynn Thorndike, 'Chiromancy', *Speculum* 40 (1965), pp. 674–706; *WIST* XIV, Catalogue 2, p. 166.

26 *WIST* XIV, Catalogue 16, p. 171. It is unattributed but occurs only with works of Adelard of Bath. Charles Burnett (see n. 2, above), pp. 135, 136, pp. 140–4; Lemay 1962, p. 8, also Lemay 1987, p.65, n.8. William of Malmesbury, *History of the Kings of England*, trans. Revd John Sharpe (London, 1815), p. 199.

27 J.V. Field, 'Astrology in Kepler's Cosmology', in Patrick Curry (ed.), pp. 143–70 (p. 144, n.3); Antonia McLean, *Humanism and the Rise of Science in Tudor England* (London, 1972), pp. 114, 115. See also F. A. Yates, *Giordano Bruno and the Hermetic Tradition* (London, 1964), p. 45.

28 John D. North, 'Some Norman Horoscopes', *WIST* XIV, pp. 147–61, Catalogue 65, p. 183. London, British Library, Royal App. 85, a MS of fragments and small leaflets of various dates. Ff. 1–2, horoscopes of s. xii, possibly drawn up by Adelard. Warner and Gilson, (BL catalogue) II, pp. 398–9.

29 Derek and Julia Parker, *A History of Astrology* (London, 1983), p.94.

30 North, *WIST* XIV, pp. 157–60. For historical background A.L. Poole, *From Domesday Book to Magna Carta*, 2nd edn (Oxford, 1955), *The Oxford History of England*, ed. Sir George Clark, 15 vols, vol. 3, pp. 161–6, 202. See ch. 10 below for association between Adelard and Henry Plantagenet.

31 *Gesta Stephani*, ed. and trans. K.R. Potter, with new Introduction and notes by R.H.C. Davis (Oxford, 1976), pp. xviii-xxxvi.

32 North, *WIST* XIV, p. 161. So far as I know

no attempt has been made to compare the handwriting here with other suggested Adelard autographs. Professor Lemay considers that 'the British Library Manuscript Sloane 2030, in the portion (ff. 83–87 actual; ff. 79–83 ancient) which contains both Adelard's *Ysagoge minor* and the *Doctrina stellarum* (*Centiloquium*) seem to me to be the original (publication) of these texts' (letter to L. Cochrane, 17 June 1984). Charles Burnett has suggested that a gloss to a copy of Boethius' *De musica* is in Adelard's hand, *WIST* XIV, p. 81. The horoscopes are illustrated *WIST* XIV, pp. 152, 153.

CHAPTER 10

The Astrolabe Treatise

Late in his career, since he cross-refers to the *Zij* of al-Khwārizmī, the translation of Euclid and to *De eodem*, Adelard wrote *De opere astrolapsus*.[1] Like his *Regule abaci*, this work is intended to instruct on the method of using a particular piece of apparatus, and it is assumed that an astrolabe is at hand. Two copies of the manuscript are well known – the Arundel MS in the British Library and the McClean MS in the Fitzwilliam Museum, Cambridge. The second incorporates an explanatory introduction and a dedication to Henry, the king's nephew or grandson (*nepos*). It is also accompanied by drawings, possibly preparatory work for an instrument to be made locally, as well as star tables which may have been included either by the author or by a later reader.[2]

Who was the dedicatee and at what date was the manuscript written? It is now generally accepted that it was dedicated to Henry Plantagenet; but the alternative suggestion should at least be mentioned. This is that it was written for Henry of Blois, nephew of King Henry I and brother of King Stephen. In 1126, when Adelard had presumably just finished the al-Khwārizmī tables, Henry of Blois was made Abbot of Glastonbury although he was still only twenty-two years of age. He later became Bishop of Winchester and having performed the coronation of Stephen supported him in the civil war which followed Henry I's death. It was thought that Adelard might have written the astrolabe treatise and dedicated it to Henry of Blois when he was made Abbot.[3]

The more likely dedicatee is Matilda's son, Henry Plantagenet, who later became Henry II. As a boy of nine he came to Bristol with his uncle, Robert of Gloucester, Matilda's half-brother. While there he could have had mathematical instruction from Adelard, although his official tutor was Master Matthew. This was in 1142. It has also been suggested that the date of the manuscript could be 1149/50 when Henry was again in England (see p. 93–4).[4]

The tone of the manuscript leads one to think that the theory concerning Henry Plantagenet is the correct one. The text is specific and didactic. It sounds more like Chaucer's treatise on the astrolabe written 200 years later for his ten-year-old son than a text for an adult already established in a church career. Whoever is the dedicatee, the manuscript begins with complimentary remarks appropriate to either:

> I know that you Henry as the King's grandson [nephew] are learned in all aspects of philosophy and appreciate that society is blessed when philosophical principles govern it ... whence it should be that you acquire not only the knowledge in Latin texts but also that which the Arabs have to teach concerning the movements of the sphere and the orbits of the stars .... Concerning the universe, therefore, and its different parts I will write in Latin what I have learned from the Arabs. You can take it for granted that the universe is not square, or rectangular but a sphere. What is said of the sphere can be said of the universe.

Vital to Adelard's explanation of the astrolabe is to establish with his reader that the universe is spherical and can be measured in terms of the circle. He is somewhat repetitious. Circular measurement was apparently unfamiliar. At the outset he spends time on definitions:

> The sphere is round and shaped like a globe with all parts equidistant from the centre. It is divided into two parts called hemispheres and then divided into quarters with the common line called the axis. The axis is a line drawn through the central point of the sphere and the two extremities are the poles. All bodies must be mobile or immobile. The body of the sphere I call the universe is mobile ...
>
> ...No parts however leave the universe, so whatever moves must do so in a circle round the centre. The poles remain immobile. One is called the Arctic, the other the Antarctic. The upper hemisphere and the one beneath our feet are separated by a line which goes from West to East. We cannot see the part beneath our feet, and this is the region which Virgil describes as belonging to departed spirits ...
>
> ...We define the universe as a sphere but it is not empty space. Within it a number of lesser bodies are contained which move round the centre. The upper hemisphere is divided into ten regions – the first or outermost has no distinguishing features to our senses; the second is known as Aplanos, or the firmament of fixed stars, because according to the Greeks, these stars do not wander; the third is the region of Saturn; the fourth of Jupiter; the fifth of Mars; the sixth of the Sun; the seventh of Venus; the eighth of Mercury; the ninth of the Moon; the tenth of the Earth.[5]

In his introduction Adelard is restating an outline of the principles which must be understood for the study of astronomy. He would not do this for someone who had completed his education, including as it would have done the subjects of the quadrivium. The text of the treatise is useful in making explicit for the general reader some of the points which are mentioned in other works. Poulle has indicated that to include a cosmology in an astrolabe treatise is most unusual, and that as regards the moon it reflects the non-Ptolemaic interpretation in al-Khwārizmī's tables.[6]

In the first two folios Adelard conveys his concept of the spherical earth at the centre of a spherical universe. The outermost circle we can see is the sky, the firmament of fixed stars – so called because they maintain the same relative positions year in and year out. As a consequence of the movement of the firmament itself philosophers have established the twenty-four hours of day and night. Since a circle has 360° each hour is represented by 15°.

If the sun's path in the sky were the same as the equinoctial circle, then day and night would be equal throughout the year. Instead the sun follows a course known as the zodiac – twelve signs each of 30° and called Aries, Taurus, Gemini, Cancer, Leo, Virgo, Libra, Scorpio, Sagittarius, Capricorn, Aquarius and Pisces. The names derive from groups of stars or constellations which follow the same course as the sun day after day and year after year. In considering the zodiac and the course of the sun one must take into account the obliquity of the ecliptic noticed by Ptolemy, according to Adelard, as 'a little more than 23°. (In fact Ptolemy's figure was 23°51′, which is considerably more.) Because of the obliquity, the ecliptic intersects the celestial equator only at the first point of Aries and the first point of Libra, called the equinoxes, one in spring and one in autumn. When this happens day and night are equal. In Cancer the sun reaches its highest point at the summer solstice (midsummer). A circle called the Tropic of Cancer is drawn on the celestial sphere to represent this fact. In Capricorn the sun is at its lowest (midwinter) and the Tropic of Capricorn is drawn at the appropriate latitude on the celestial sphere. The first point of Aries and the first point of Libra divide the zodiac and also the equator into two equal parts. For one half of the year the sun is above the celestial equator at midday and for the other half it is below.

Poulle's view is that there would normally be three possible objectives in writing an astrolabe treatise: to explain stereographic projection; to describe how an astrolabe is constructed; and to give information on how it is used. Adelard's intention is to discuss how the astrolabe is used. His introductory cosmology provides a background for the stereographic projection which he does not attempt to explain. Hermann of Carinthia translated Ptolemy's *Planisphere* in 1143 so information was available if Adelard was writing in 1149. Adelard does stress the fact that with the astrolabe a three-dimensional sphere is illustrated on a two-dimensional plane. He remarks on the genius of Ptolemy in establishing that observations of the sphere can be made by using rules which apply to circles. This is an indirect reference to Hermann's translation. He makes clear how various circles projected of the sphere are used for astronomical observation. The ecliptic is another name for the zodiac, and is so called because eclipses of the sun and moon take place along its course. It is a great circle, the description given to any circle on the surface of a sphere whose plane passes through the centre of that sphere. A celestial great circle has the same circumference as the celestial equator. Adelard explains the great circles called colures which pass through the equinoxes and the solstices. The one through the equinoxes joins Aries and Libra and is known as the east-west line. (Adelard uses the term 'orion', presumably for [*linea*] *orientalis*.) The line through the solstices is the meridian. When these two great circles of the sphere are drawn on the plane of the astrolabe, they are the central diameters and appear as straight lines.[7]

Adelard does not explain why some great circles become straight lines on the plane. It is because they are circles through the source of projection and a special case, that is, circles of infinite radii. Other circles of the sphere project as circles on the plane, and an angle at any point on the sphere between two circles is equal to the angle at the point of intersection of the projected curves. These mathematical properties of stereographic projection make the astrolabe possible. Astrolabists rarely bothered to explore the theoretical background. They were content just to construct and use the instrument. The theoretical elements of stereographic projection are thought to have been first understood by Hipparchus in the second century BC. The source of the projection for an astrolabe dealing with the northern hemisphere is the celestial South Pole.[8]

Certain signs are always in opposition to each other and six months apart in the annual circuit of the sun. Within the 30° of each sign the numeration of opposite signs will be the same. Gemini 17 will always be opposite to Sagittarius 17. In the course of a year it takes six months for the positions to reverse. In the course of twenty-four hours it is only the difference between noon and midnight. Adelard's explanation is preparing the ground for points he discusses in detail later. A diagram is included in the McClean MS.[9]

Adelard states that he has sufficiently explained the planetary system in his book on the Arabic *zij* and he will not repeat the information. He gives a brief summary of the thirty-year cycle of Saturn. He has already indicated that the planets follow courses which differ from the ecliptic and appear sometimes stationary or retrograde. Another diagram in the McClean MS is intended to show how the cycle of Saturn relates to the ecliptic.[10]

This is followed by Adelard's explanation of *clima* ('climate'). In Adelard's time the term referred to the 'supposed slope or inclination of the earth and sky from the equator to the poles' and meant 'the zone or region of the earth occupying a particular elevation on this slope, that is, lying in the same parallel of latitude'. In Ptolemy's time seven 'climates' were reckoned, supposed to be presided over by the seven planets. Later there were twenty-four between the equator and the pole, each representing the increase of an extra half hour of daylight at the time of the summer solstice.[11]

This helps to clarify the significance of Arin's supposed geographical location. As Adelard puts it, since all measurements derive from circles of 360° the philosophers had decided that there must be some point on the earth's surface where measurements commenced and that Arin was to be the place. They limited their considerations to one half of the northern hemisphere. They placed Arin on the equator so as to make it 90° from the North Pole and midway between east and west. 'Climate plate' is another term for the latitude plate inserted in an astrolabe to enable an observer to calculate the altitude and position of heavenly bodies in relation to his own position on the earth's surface. Adelard uses his birthplace, Bath, as his example. Bath is 52° N. of Arin. This means that to use an astrolabe in Bath one must have a 'climate plate' for 52°.

Adelard now considers that the preliminaries are complete and he starts his detailed description of the astrolabe and its uses. It is at this point that the Arundel MS commences.[12] Only the McClean MS contains the introductory dedication and cosmology.

Poulle points out that Adelard's description of the astrolabe and its uses is 'topographical', since he comments on function as he examines each section of the astrolabe in turn. Adelard begins by saying that it is a flat, circular instrument made of metal which represents the universe, and by using it according to the design engraved upon it one can make scientific observations based on angles. He describes the back first (see Plate 9).

The outer rim (limb) of the instrument is divided into marked angles of 360° so an angle of elevation can be taken. It is also subdivided into calendar months with days co-ordinated with signs of the zodiac each with 30°. In the centre is a pointer called the alidade which is used for sighting along a diameter of the circle when the astrolabe is suspended vertically. A unique feature on which Poulle particularly comments is Adelard's reference to the fact that the alidade could be a tube, in effect like the *fistula* used by Gerbert in the spherical astrolabe which he designed.[13] Normally the alidade is a narrow revolving strip of metal on which two sighting vanes with pinhead-size holes are mounted. It is through these holes that one can sight a star directly or take the altitude of the sun indirectly.

Adelard begins his summary of the astrolabe's working functions with its usefulness as a surveying instrument. This is quite straightforward. Practical geometry is involved. There are quadrants called shadow squares with one side as Umbra Recta and one as Umbra Versa (Indirecta) on the back of the instrument which can be used with the alidade. Incorporated in both Arundel and McClean MSS are sketches which demonstrate how, for example, the height of a tower can be measured. The figures used are roman numerals, not arabic.[14] (See Figure 4.)

Both the *Zij* and the translation of Euclid's *Elements* are mentioned in the course of Adelard's discussion of how the astrolabe is used for astronomy and navigation. He explains the procedure for establishing one's latitude, and defines the terms *chemeli* ('northern') and *genubia* ('southern'). He states that one's latitude can be determined at the first point of Aries and the first point of Libra by taking the altitude of the sun at midday and subtracting it from 90°. This method of observation assumes that one's zenith, directly above one's head, is 90° from a tangent to the earth's surface at one's feet. Another method of establishing latitude with the astrolabe is to use it at night. To do this a circumpolar star is selected that never sets and its elevation is measured above the horizon where it cuts the meridian in its circular course around the pole, establishing the maximum and the minimum angles of elevation. The diameter of this circle is then measured, and the pole will be exactly halfway between the highest and the lowest points.[15]

Adelard now comes to the complicated problem of establishing longitude which was also mentioned in connection with the *Zij*. It relates to the problem of calculating time. Modern man of course knows that the earth is moving around the sun and simultaneously making a complete revolution itself every twenty-four hours. Each day the sun is due south minute by minute for every place on the earth's surface. When the sun is at its highest point for the day, due south on the meridian, it is noon, local solar time. To establish longitude, however, it is essential to calculate differences in local time, measured from a prime meridian. What is required is that a celestial event, like a lunar eclipse, which is visible simultaneously from all parts of a hemisphere of the earth's surface, should be recorded by two observers in terms of local time. If the true longitude of one station is known the other can be established.

Adelard apparently did not fully understand the problem. He suggests it is possible to use an eclipse of the sun for this purpose also, that to know the longitude of Bath one must know at what time an eclipse of the sun or moon happens in Arin, and having noted the

Diagrams of how the astrolabe is used for surveying:

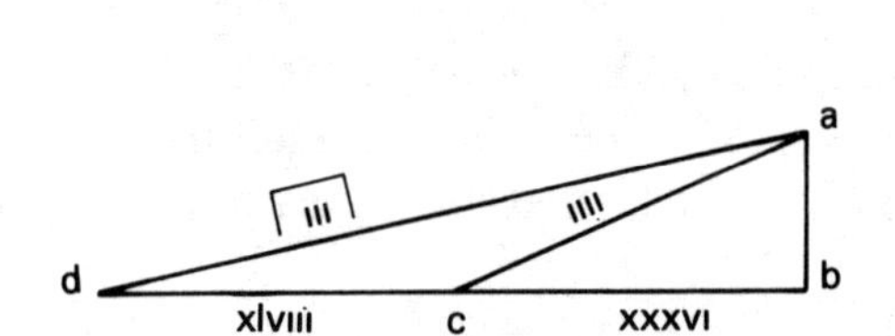

A1 A measurement when access is impeded, Umbra Versa, based on Arundel 377, f. 70r., British Library

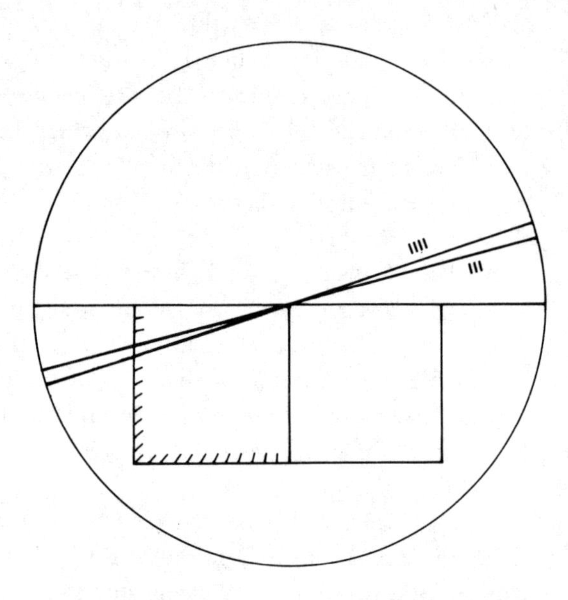

A2 Position of alidade on astrolabe, author's diagram

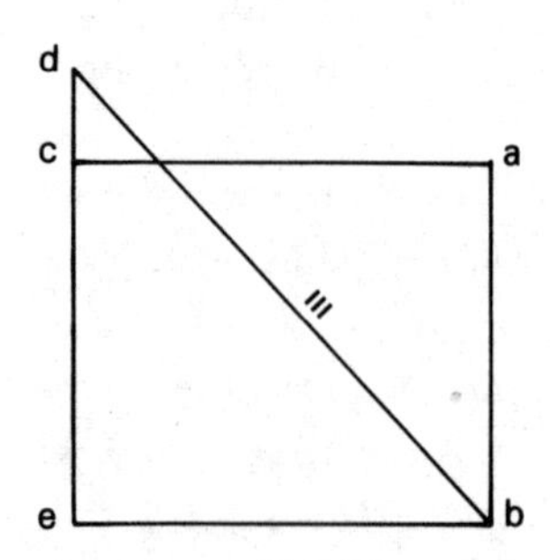

B1 Measuring the depth of a pit, Umbra Recta, based on Arundel 377, f. 70r., British Library

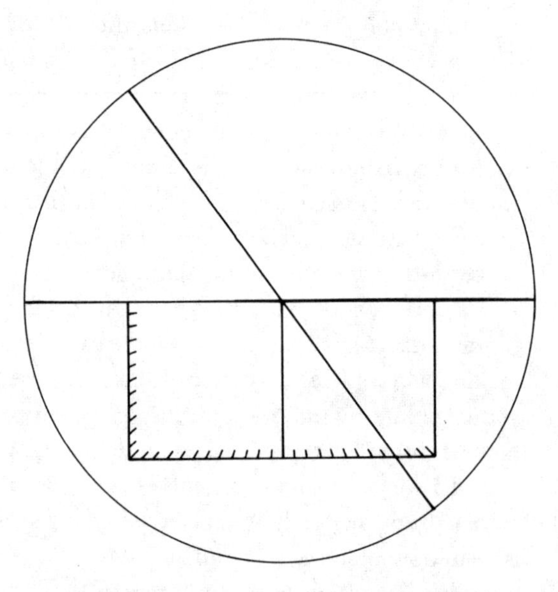

B2 Position of alidade on astrolabe, author's diagram

4 The astrolabe used for surveying, based on BL Arundel 377, f. 70r., with added diagrammatic explanation.

time at which the same event is observed in Bath the difference in hours will give the difference in degrees of longitude, allowing 15° for each hour.[16]

Adelard is also wildly out in his calculation: he gives Bath's longitude as 45° W. of Arin – an event on the meridian at Arin will be seen at Bath at the third hour, three hours difference, hence 45°. Since Bath is 78° W. of Ujjain, the modern Indian city on the site of Arin, the error is considerable, even taking into account the variations discussed in Chapter 8. It is interesting that Petrus Alfonsi deals very confidently with theoretical calculations concerning Arin in his *Dialogus*. It may be that the theoretical suppositions led to confusion. Adelard was perhaps conscious of the fact that his method left something to be desired. He also mentions that one can use information about the moon's appearance on the meridian, comparing lunar tables for different localities as a method of calculating longitude. This was understood by Gerard of Cremona writing a brief treatise on astronomy later in the twelfth century.[17]

The remainder of the treatise is devoted to the front or face of the astrolabe (see Plates 8 and 10). Adelard deals now with how to use the instrument for other calculations like telling local time, as well as for astrology. It is clear from Adelard's comments that he is describing the type of astrolabe commonly utilised in the West. The British Museum has one of the earliest European astrolabes, *c.* AD 1195. It differs from eastern astrolabes in the use of Latin terms for the signs of the zodiac and for the calendar months. It is interesting in having a form of arabic numerals to calculate the degrees around the circumference. Earlier eastern astrolabes used abjad numerals, alphabetical symbols for numbers.

A Hispano-Moorish astrolabe in the National Museums of Scotland (Royal Scottish Museum) bears the following inscription on the back in Kufic script: 'Work of Muhammed b. as-Saffâr in Cordova in the year seventeen and four hundred' (i.e. AD 1026). The rete is a thirteenth-century replacement made to allow for precession so that the astrolabe could continue in use. The rete shows Syrian influence in the design with severe dagger-shaped star pointers.[18]

It was usual for the front of an astrolabe, called the 'Mother' or the 'Womb', to be hollowed out so that the circular 'climate plate' appropriate to a particular latitude could be inserted. This was fixed in place by a pin, through which was pushed a peg with a head shaped like that of a horse. The outer rim of the astrolabe's front was also divided into 360°. Just within this came the 'climate plate', the outermost circle representing the Tropic of Capricorn. The next circle was the celestial equator, and then came the Tropic of Cancer. The celestial North Pole was in the centre. The two diameters at right angles were the meridian and the equinoctial colure. A series of concentric circles, called almucantars, represented degrees of latitude from the horizon to the observer's zenith. All these had to be correct for a particular latitude. This was why a number of different plates were required.[19]

Adelard lays great emphasis on remembering how important it is in calculating the position of the sun, moon or planets to add or subtract the obliquity of the ecliptic according to the time of year.[20]

Above the climate plate the revolving rete shows the signs of the zodiac each divided into 30° and also important stars like Sirius, Aldebaran, Capella, Vega, Altair and Spica.

The star pointers of a well-designed instrument touch the climate plate with great accuracy. Lines of azimuth which give compass bearings around the horizon provide an additional co-ordinate when locating a star. The eastern horizon is on the left, south at the top of the straight line representing the meridian, and west on the right.

The planispheric astrolabe is a quite remarkable instrument. Because it can be set to show the positions of heavenly bodies at different times of day or night, on different dates and for different latitudes, it can be used as a sort of computer to solve problems which depend on these movements, such as the length of day or night at any particular time.[21]

North has summarised the astrolabe's role:

> The astrolabe was the most widely used astronomical instrument in the Middle Ages. It originated in antiquity and was still not uncommon in the 17th century. One purpose of the instrument was observational: it was employed for finding the angle of the sun, the moon, the planets or the stars above the horizon or from the zenith. It could also be used for determining the height of mountains and towers or the depth of wells and for surveying in general. Far more important, however, was the astrolabe's value as an auxiliary computing device. It enabled the astronomer to work out the position of the sun and principal stars with respect to the meridian as well as the horizon, to find his geographical latitude and the direction of true north (even by day, when the stars were not visible), and it allowed him to indulge in such prestigious and lucrative duties as the casting of horoscopes. Above all in the days before reliable clocks were commonly available, the astrolabe provided its owner with a means of telling time by day or by night, as long as the sun or some recognizable star marked on the instrument was visible ....
>
> ...The chief purpose of the astrolabe was for telling the time. First the altitude of the sun or a star was found by employing it as an observing instrument. Then assuming that the observer knew where the sun or the star was on the rete, the rete was revolved until that point coincided with the almucantar for the appropriate altitude. (It is assumed that the observer knew which climate to choose for his latitude and on which side of the meridian line the object fell.) The refraction of the atmosphere which changes the apparent position of objects in the sky, and which is greater the nearer they are to the horizon, was ignored. The sun's approximate position on the ecliptic for any day of the year is found from the calendar scale on the back of the astrolabe.[22]

Poulle has emphasised his view that astrolabes were far more useful as calculating or instructional instruments than they were for observation. This was one reason why he was so interested in Adelard's suggestion that the alidade might take the form of a tube. A tube is easier to sight through than two pinholes. The calculation of angles on the surface of an astrolabe was also easy to explain with the instrument to hand. In addition to the alidade on the back there is a 'rule' or 'label' which revolves on the front to assist calculations.

Adelard completes his treatise with a detailed explanation of how the astrolabe is used for astrology, beginning with how to establish the ascendent and divide the celestial sphere into the twelve houses. He uses the method of setting up celestial houses which

North calls 'standard'. Once the rete has been set for the purpose of casting a horoscope a point on the ecliptic becomes the ascendent rising in the East. If the rule or label is laid so that its edge passes through that ascendent it will make a certain angle with the meridian line – the vertical line on the plate. Mid-heaven is where the meridian meets the ecliptic and that is a cardinal point of the division into houses. Poulle draws attention to the fact that although there were four well-known methods of dividing the firmament into houses in the twelfth century Adelard and his colleagues favoured a particular one.[23]

The influences which affect horoscopes are the result of the houses in which sun, moon and planets happen to be at the given moment. The houses themselves are responsible for different sorts of influence related to life, business, brothers, parents, children, illness, marriage, death, travel, honours, friends and enemies. Having established with the astrolabe the sign of the zodiac and the time of day, one can discover the position of the sun and which constellations will be significant. Tables are then consulted for further information on the moon and the various planets. The 'aspect', or geometrical relationship planets have to one another, is very significant in horoscopes. *Radiatio* or radiation was similarly important. Poulle is particularly impressed by the fact that Adelard describes how to calculate *radiationes.* These are astrological rays which Ptolemy had postulated of the planets. (The planets regarded others ahead of them in the order of the signs and projected rays back at those following them.) Ptolemy calculated *radiationes* as 'right ascensional aspects' (to quote North). Simple aspects calculated in ecliptic longitudes should offer no problems to anyone and scarcely merit the use of an astrolabe. Radiation was sent by a planet to right or left, when the corresponding points on the equator were 60, 90 or 120° separation. It was a simple matter to use the astrolabe to find the arcs involved.[24]

Certain features common to most astrolabes are not incorporated in the sketches which accompany the McClean manuscript, nor are they made clear from the text. The mathematical techniques for establishing the centres from which almucantars (showing degrees of celestial latitude) must be drawn are not discussed. Further information would be required to construct an instrument. The latitudes in the diagrams are consistent with a dedication to Henry Plantagenet since the climate plates are for southern England and northern France.[25]

Adelard's treatise has become the subject of much more interest since the suggestion was put forward that he had been acting as a court astrologer. The occasion for which he wrote the treatise is thought now to have been during a visit to England by Henry in 1149/50 when he was knighted by David of Scotland. This was shortly before Henry's coming of age. It would fit well with the fact that one of the group of ten horoscopes was produced to mark the occasion of a meeting between a master and a former pupil.[26] This link between Adelard's final work and the reign of Henry II rounds up his quite extraordinary career in an entirely gratifying fashion.

Notes

1 Adelard of Bath, *De opere astrolapsus*, *WIST* XIV, Catalogue 10, Cambridge, Fitzwilliam Museum, McClean MS 165, ff. 81–8; British Library, Arundel MS 377, ff. 69–74; Emmanuel Poulle, 'Le traité de l'astrolabe d'Adélard de Bath', *WIST* XIV, pp. 119–32.

2 McClean, ff. 87,88. Drawings are reproduced in Louise Cochrane, 'Adelard of Bath and the Astrolabe', *PSANHS* 124 (1980), pp. 147–50. Poulle (see n.1) refers to the table of stars, p. 124. See also comment in an Addendum, by P. Kunitzsch, *WIST* XIV, p. 131.

3 Haskins 1927, p. 29.

4 Haskins 1927, p. 29. The suggestion that the date might be 1149/50 was first proposed by B.G. Dickey, in 'Adelard of Bath, An Examination Based on Heretofore Unexamined Manuscripts', an unpublished Ph.D. dissertation, University of Toronto, 1982, pp. 7–11, and cited in North, 'Some Norman Horoscopes', *WIST* XIV, p. 159. North supports the view that Henry Plantagenet, not Henry of Blois, is the dedicatee.

5 McClean, f. 81r.

6 Poulle, *WIST* XIV, pp. 120, 121.

7 McClean, f. 81v. Comment on Ptolemy, f. 83r.

8 Ron Thomson, 'Jordanus de Nemore and the Mathematics of Astrolabes', *Pontifical Institute of Mediaeval Studies* 39 (Toronto, 1978); Thomson records in an appendix a Russian view that the early Arab astronomer al-Farghāni had a proof which eluded Western scientists until much later. O. Neugebauer, 'Early History of the Astrolabe', *Isis* 40 (1949), p. 246. See also John D. North, 'The Astrolabe', *Scientific American* 230, no.1 (January 1974), pp. 96–106, repub. in John D. North, *Stars, Minds and Fate, Essays in Ancient and Medieval Cosmology* (London and Ronceverte, West Virginia, 1989), ch. 14, pp. 211–20; John D. North, *Chaucer's Universe* (London, 1988); D.W. Waters, *The Planispheric Astrolabe*, National Maritime Museum (Greenwich, 1976).

9 McClean, f. 81v.

10 McClean, f. 82r.

11 McClean, f. 82v. Definition of 'climate', *OED*, vol. 1, p. 435.

12 McClean, f. 83r. (Arundel 377, f. 69, begins at the point where the instrument is described.)

13 Poulle, *WIST* XIV, p. 123; McClean, f. 83r. (description of alidade).

14 McClean, ff. 83v., 84 r. There are similar illustrations in Arundel 69v., 70r.

15 McClean, f. 84v. J.K. Wright, 'Notes on the knowledge of latitudes and longitudes in the middle ages', *Isis* 5, part 1 (Brussels, 1923), pp. 76–98 (78–80).

16 McClean, ff. 84v., 85r.; Arundel 70v., 71r.; Poulle, *WIST* XIV, pp. 125, 126.

17 Wright 1923, pp. 81–91; Petrus Alfonsi, *Dialogi cum Judaeo*, Migne, *Patrologia Latina* 157, cols 535 ff.; Patrick Moore, *The Observer's Book of Astronomy* (London, 1978), p. 146, for explanation of lunar eclipses. The Arin theory survived until the time of Christopher Columbus – see D.M. Dunlop, *Arab Civilization to A.D. 1500* (London, 1971), p. 156.

18 British Museum Catalogue, European Scientific Instruments 323, Astrolabe Plate XLIX, European, *c.* 1195, pp. 110–12; National Museums of Scotland (Royal Scottish Museum) Hispano-Moorish astrolabe: Muhammad b. as-Saffâr, Cordova, AD 1026, presented by J.H. Farr. I am indebted to Professor David King for information on the rete of the Hispano-Moorish astrolabe.

19 For background information on astrolabes the author particularly recommends North 1974 (repub.1989); North, 1988; Waters 1976; Geoffrey Chaucer, 'Treatise on the Astrolabe', ed. W.W. Skeat, *Early English Text Society* 16 (1872) for Skeat's Introduction and diagrams.

20 McClean, f. 86r; Arundel 72 v.

21 Waters, p. 8.

22 North 1974, pp. 96, 105; repub.1989, p. 211.

23 John D. North, *Horoscopes and History*, *WIST* XIII (1986), p. 59; North 1988, pp. 65, 103; Poulle, *WIST* XIV, p. 128.

24 McClean, 86v.; Poulle, *WIST* XIV, p. 130; North 1988, pp. 220–9.

25 See n. 2, above.

26 See n. 4, above; also ch. 9, p. 93–4.

Conclusion

How do we sum up Adelard's career? His intellectual development reflects trends now well established as representative of the history of ideas in the twelfth century. He was the beneficiary with regard to some; he was the initiator of others. The early work, *De eodem et diverso*, is a comprehensive survey of what was taught in outstanding European cathedral schools at the time, and its style reflects a desire to achieve a literary standard worthy of those who introduced him to the works he wished to emulate.

Adelard's early mathematical promise was undoubtedly developed by good teaching. This is demonstrated by the treatise on the abacus and then more importantly in the translation of Euclid's geometry from Arabic into Latin and (possibly) his algorism. Without a sound early mathematical education this would have been inconceivable. A new emphasis on the quadrivium was already establishing itself and Adelard made the most of his opportunities. Mathematics is not a subject one can teach oneself easily from books. The anonymous teacher who left Adelard to contemplate the stars on the banks of the Loire must be credited with helping to provide the foundation on which Adelard was later able to build so successfully. So also must be the Bishop of Syracuse.

Adelard's interests developed in the direction of cosmology and astronomy but not theology. He began as a neo-Platonist, emerging from Tours and Laon with a firm grounding in the liberal arts and the attitude of mind we now call medieval humanism. He chose at first to write allegorically and attempted to combine poetry with prose. This literary form is thought to have had special significance. The personification of Philosophia and Philocosmia was used as a means of conveying deeper meanings associated with a scientific explanation of the universe. Adelard's neo-Platonism yielded to an increasing interest in Aristotelianism. This was reflected in *Quaestiones naturales*, although Adelard did not appear to have consulted any of Aristotle's works directly in Arabic as he did with Euclid. He chose to translate al-Khwārizmī and astrological works rather than general philosophical ones. But with *Quaestiones* he made his own contribution to the gradual acceptance of Aristotle's 'new logic'. He also adopted a new literary form. Here he did not use allegory but a direct method of question and answer. The presentation could be taken as a means of ensuring that conventional ideas were expressed by his nephew and that unusual or new opinions were those of his Arab teachers with whom he did not necessarily agree. This form of protective colouring was not required, however, for the simple and straightforward treatise on the care of falcons in which Adelard again uses his nephew as a literary device.

The translation of Euclid was unquestionably of primary importance. Although obviously not the result of original thought, it demonstrated a higher level of mathematical competence than the limited versions previously available. Not only to understand the Arabic but to explain the mathematics must have required many months, even years, of concentrated effort.

The real enigma with regard to Adelard's life and work is that having meticulously explained the importance of the exact positions of heavenly bodies *vis à vis* the earth's surface and how to locate them, he blatantly disregarded the difference in latitude between Bath and Cordova when casting horoscopes and made no real attempt by observation or otherwise to produce accurate astronomical tables for his own use. The only explanation of this which seems feasible is that once he began translating Euclid's geometry, mathematics again overtook his other interests both as translator and as teacher, and not until late in life was he required to act himself as an astrologer.[1] By this time younger men had taken up some of the work he had initiated – men like Robert of Chester, Daniel of Morley and Roger of Hereford in England. Hereford was developing as an important centre of scientific studies. Only gradually did the increasing emphasis on astrology become a source of serious concern to Church leaders.

One gets the impression that Adelard was indebted to others like himself in his own generation, that he would not have arrived at his views working completely alone. What we must be grateful for is that among a number of scholars who brought scientific information to the attention of the Latin West Adelard took the initiative not only in translating Arabic works but in recording their usefulness and developing the reasoning on which they were based. It is Adelard as author to whom we must be grateful as well as to Adelard the scholar.

Notes

1 North, *WIST* XIV, pp. 159–61, refers to Horoscope C (1123) as probably having been cast for Henry I; the next date is 1135, thereafter 1150, 1151 (six horoscopes) and the last 1160.

Bibliography

I Historical records

Charters witnessed for Bishop John of Bath: W. Hunt, *Two Chartularies of the Priory of St Peter at Bath*, Somerset Record Society, 7 (1893), 2 vols, vol. 1, nos 34 (no date), 41 (1100), 53 (1106), 54 (1108); vol. 2, no. 844, another version of no. 34

DB, f. 89 r.-v.; Exon Domesday, pp. 145–7

Historical Manuscripts Commission, *Calendar of the Manuscripts of the Dean and Chapter of Wells*, 2 vols (1907, 1914), Diploma of King William I, 1068, granting at the request of Giso, bishop (of Wells) thirty hides of land at Banwell (Somerset), vol. 1, p. 431

MS London, British Library, Harley 358, ed. W. Dugdale, *Monasticon anglicanum*, 2nd edn (London, 1819), vol. 2, p. 268, same as no. 53, Hunt (above) except for 'Walterus vicecomes add Hosatus'

Pipe Roll, 31 Henry I (1130), ed. J. Hunter, p. 22

II Manuscript sources

Centiloquium Ptolemei
'Doctrina stellarum ex te et illis ... ', British Library, Sloane MS 2030, ff. 82, 83

Ciromantia
'Linee naturales tres sunt in planitie ... ', British Library, Sloane MS 2030, f. 121

De opere astrolapsus
1. 'Erit igitur ut astrolabium de mag. Adelard', British Library, Arundel MS 377, ff. 69–74.
2. 'Incipit Libellus Magistri Alardi Bathoniensis de opere astrolapsus', Fitzwilliam Museum, McClean MS 165, ff. 81–8

Elementa, Euclid
'Artis Geometrice per Adelardum Bathoniensem ex arabica lingus in latine translate', British Library, Old Royal and Kings MS, 15A, XXVII (only one of numerous manuscripts – see below)

Horoscopes (twelfth century, possibly drawn up by Adelard)
British Library, Royal App. 85, MS of fragments and small leaflets of various dates, ff. 1–2

Isagoge minor, Abū Ma'shar
'Ysagoga minor Japharis matematici in astronomiam per Adhelardum bathoniensem ex arabico sumpta', British Library, Sloane MS 2030, ff. 83–6

Liber prestigiorum Thebidis
Lyon, Bibliothèque Municipale, 328; Adelard's translation of a work by Thābit b. Qurra on magical sigils

Regule abaci
Leiden, Bibliotheek der Rijksuniversiteit, Scaliger 1

For detailed references to other available manuscripts, texts and commentaries:

Burnett, Charles (ed.), *WIST* XIV, Catalogue, pp. 163–196

Thorndike, Lynn, and Kibre, Pearl, *A Catalogue of Incipits of Mediaeval Scientific Writings in Latin*, Mediaeval Academy of America (London, 1963)

III Editions of texts

Adelard as author or translator
Abū Ma'shar, *Abū Ma'šār, The Abbreviation of the Introduction to Astrology, together with the Medieval Latin Translation of Adelard of Bath*, ed. and trans.

Charles Burnett, Keiji Yamamoto, Michio Yano (Leiden, 1994)

Abū Ma'shar, *Ysagoge minor* ('Shorter Introduction to Astronomy'), trans. Adelard of Bath, Oxford, Bodleian MS Digby 68, f. 116; *see* Richard Lemay, *Abū Ma'shar and Latin Aristotelianism in the 12th Century* (Beirut, 1962), p.355

De cura accipitrum ('The Care of Falcons'), ed. A.E.H. Swaen (University of Amsterdam, 1937)

De eodem et diverso, ed. H. Willner, 'Des Adelard von Bath Traktat', *Beiträge zur Geschichte der Philosophie (und Theologie) des Mittelalters* 4, 1, (Münster, 1903)

Euclid, *Elementa*, trans. Adelard of Bath
M. Clagett, 'The Medieval Latin Translations from the Arabic of the *Elements* with Special Emphasis on the Versions of Adelard of Bath', *Isis* 45 (1953), pp. 16–42, first established the categories known as Adelard I, II, III

H.L.L. Busard, *The First Latin Translation of Euclid's Elements Commonly Ascribed to Adelard of Bath*, Pontifical Institute of Mediaeval Studies and Texts 64 (Toronto, 1983), an edition of Adelard I

H.L.L. Busard and Menso Folkerts are preparing a critical edition of Adelard II

The preface of Adelard III is edited by M. Clagett, 'King Alfred and the *Elements* of Euclid', *Isis* 45 (1954), pp. 69–77

Incipit Ezich Elkaurezmi per Athelardum Bathoniensem ex Arabico sumptus (Tables of Muhammed ibn Musa al-Khwārizmī)
H. Suter, A. Björnbo and R.O. Besthorn (eds), *Die astronomischen Tafeln des Muhammed ibn Musa al-Khwārizmī in der Bearbeitung des Maslama ibn Ahmed al-Madjrītī und der latein Überssetzung des Athelard von Bath auf Grund der Vorarbeiten von A. Bjornbo und R.O. Besthorn*, Det Kongelige Danske Videnskabernes Selskab, histor.-filosof. Skrifter 7, Raekke, Afd. 3 (Copenhagen, 1914)

O. Neugebauer (ed. and trans.), *The Astronomical Tables of al-Khwārizmī*, Det Kongelige Danske Videnskabernes Selskab, histor.-filosof. Skrifter, 4, 2 (Copenhagen, 1962)

Quaestiones naturales
H. Gollancz (ed. and trans.), *Dodi ve Nechdi.* The work of Berachya Hanakdan, edited from MSS at Munich and Oxford, to which is added the first English translation of Adelard of Bath's *Quaestiones naturales* (London, 1920). The book contains Sir Hermann Gollancz's translations of two Hebrew versions of Adelard's text as well as a translation into English from the first printed edition (Latin, 1480)

Martin Müller (ed.), 'Die Quaestiones Naturales des Adelardus von Bath', *Beiträge zur Geschichte der Philosophie (und Theologie) des Mittelalters* 31, 2 (Münster, 1934)

Regule abaci
B. Boncompagni (ed.), 'Intorno ad uno scritto inedito de Adelardo di Bath intitolato "Regule abaci"', *Bulletino de Bibliografia e di Storia de le Scienze Matematiche* 14 (1881), pp. 1–134

Adelard(?) as translator of al-Khwārizmī's algorism (arithmetic)

Liber ysagogarum Alchorismi
'Liber ysagogarum Alchorismi in artem astronomicam a Magistro A compositus', a work in five books
A. Allard (ed.), *Les plus anciennes versions latines du XIIe siècle issues de l'Arithmétique d'al-Khwārizmī*, Centre d'ecdotique et d'histoire des sciences, Université de Louvain (Louvain, 1975), pp. 93–145

M. Curtze (ed.), first three books, 'Über eine Algorismus-schrift des XII. Jahrhunderts', *Abhandlungen zur Geschichte der Mathematik* 8 (1898), pp. 1–27

De numero Indorum
'Dixit algorizmi', Cambridge University Library MS Ii
B. Boncompagni (ed.), *Trattati d'Aritmetica* (Rome, 1857)

A.P. Juschkewitsch (ed.), 'Über ein Werk des Abdallah Muhammed ibn Musa al Huwarazmi al-Magusi zur Arithmetik der Inder', *Schriftenreihe für Geschichte der Naturwissenschaften, Technik und Medizin, Beiträge zur Geschichte der Naturwissenschaften, Technik und Medizin herausgegeben zum 60. Geburtstag Gerhard Harigs*, ed. I. Strube and H. Wussing (Leipzig, 1964), pp. 21–63

K. Vogel (ed.), *Mohammed ibn Musa Alchwarizmi's Algorismus: das fruheste Lehrbuch zum Rechnen mit Indischen Ziffern* (Aalen, 1963), pp. 1–63

Adelard(?) in possession of a text, possible contributor

Ergaphalau
Charles Burnett (ed.), 'Adelard, Ergaphalau and the Science of the Stars', *WIST* XIV, pp. 133–45

Mappae clavicula, attrib. to Adelard in the Contents of British Library, Royal and Kings 15C IV, which no longer includes the work

Sir Thomas Phillipps and Albert Way (eds), '*Mappae Clavicula*, A Treatise on the Preparation of Pigments during the Middle Ages', *Archaeologia* 32 (1847), pp. 183–244

C.S. Smith and J.G. Hawthorne (eds), '*Mappae Clavicula*, A Little Key to the World of Medieval Techniques', *TAPS* 64 (Philadelphia, 1974)

Adelard(?) suggested as possible translator
De differentia spiritus et anime, Qustā b. Lūqā, *see WIST* XIV Catalogue 5, p. 167

IV Editions of related texts

Alfonsi, Petrus (Alphonso, Pedro), 'Dialogi cum Judaeo', J.P. Migne, *Patrologia Latina* 157, cols 535–672

Alfonsi, Petrus, *A Scholar's Guide, Disciplina Clericalis*, ed. D. Labouderie (Paris, 1824) repr. J.P. Migne, *Patralogia Latina*, 157, cols. 671–706, trans. J.R. Jones and J.E. Keller (Toronto, 1969); ed. and trans. by Eberhard Hermes, trans. into English by P.R. Quarrie (London, 1977)

Alfonsi, Petrus, *Letter to the Peripatetics of France*, London, BL, Arundel 270, an introduction intended as a preface to a work on astronomy accompanying a version of the *Tables of al-Khwārizmī*. *See* Walcher of Malvern and Petrus Alfonsi

Boethius, *Tractatus de Consolatione Philosophiae*, Loeb Classical Library (London, 1973). For 'Boethius' *Geometry II*, *see* Menso Folkerts, Section V below

Euclid, *Elements*, 'The Thirteen Books of Euclid's *Elements*', 'The Works of Archimedes', trans. Sir Thomas L. Heath, *Great Books of the Western World*, ed.-in-chief, Robert M. Hutchins, vol. 11 (Chicago and London, 1952)

Fitznigel, Richard, *Dialogus de Scaccario, The Course of the Exchequer by Richard, Son of Nigel*, ed. and trans. by the late Charles Johnson with corrections by F.E.L. Carter and D.E. Greenway (Oxford, 1983)

Gerberti, postea Sylvester II, Papae, *Opera Mathematica*, ed. Nicholas Bubnov (Berlin, 1899)

Gerbert, *Letters*, trans. with introduction by H.P. Lattin (New York, 1961)

Gesta Stephani, ed. and trans. K.R. Potter, with new Introduction and notes R.H.C. Davis (Oxford, 1976)

Hermann of Carinthia, *De essentiis*, ed. Charles Burnett (Leiden, 1982)

Plato, *Timaeus*, ed. and trans. F.M. Cornford, *Plato's Cosmology, the Timaeus of Plato, translated with a running commentary* (London, New York, 1937); Desmond Lee (ed.), *Timaeus and Critias* (Harmondsworth, 1965); A.E. Taylor (ed.), *A Commentary on Plato's Timaeus* (Oxford, 1928)

Ptolemy, Claudius, *The Almagest*, ed. and trans. G. Taliaferro, *Great Books of the Western World*, ed.-in-chief, Robert M. Hutchins, vol. 16 (Chicago, London, 1963); ed. and trans. G.J. Toomer (London, 1984)

Ptolemy, Claudius, *Tetrabiblos*, ed. and trans. F.E. Robbins, Loeb Classical Library (London and Cambridge, Mass, 1940)

'Ptolemy', *Centiloquium*, *see* Richard Lemay, 'Origin and Success of the *Kitāb Thamara* of Abū Ja'far Ahmad ibn Yūsuf ibn Ibrahīm', *Proceedings of the First International Symposium for the History of Arabic Science, April 5–12, 1976*, 2 vols (Aleppo, 1978), vol. 2, pp. 91–107. *See also* Ashmand, J.M., Section V below

Ralph (Radulphus) of Laon, ed. A. Nagl, 'Der Aritmetische Traktat des Radulph von Laon', *Zeitschrifte für Mathematik* 34 (1889), Supplement (1890) hist.-liter Abt. pp. 85–133; e codice Parisiense latino no 15120

Walcher of Malvern and Petrus Alfonsi, *De dracone*, ed. J.M. Vallicrosa, 'La aportación astronmica de Pedro Alfonso', *Sefarad* 3 (1943), pp. 87–97, Oxford, BL, Auct.F.1.9; *WIST* XIV Catalogue 6

V Books and journals

Adamson, Robert, 'Adelard of Bath', *DNB*, p. 12

Afnan, Soheil M., *Avicenna, His Life and Works* (London, 1958)

Alfonsi, Petrus, *see* Section IV above

Allard, A., 'L'époque d'Adélard et les chiffres arabes dans les manuscrits latins d'arithmétique', *WIST* XIV, pp. 37–43. *See also* Section III above, *Liber ysagogarum Alchorismi*

Anderson, R.G.W., *The Mariner's Astrolabe*, Royal Scottish Museum (Edinburgh, 1972)

Arnaldez, R., and Massignon, L., 'Arabic Science', in René Taton (ed.), *Ancient and Medieval*

Science, A General History of the Sciences, 4 vols, (London, 1963), vol. 1, pp. 385–421

Ashmand, J.M. (ed.), Ptolemy, Claudius (trans. from Greek text of Proclus), *Ptolemy's Tetrabiblos or Quadripartite, being Four Books of the Influence of the Stars with a Preface and Explanatory Notes and an Appendix containing Extracts from the* Almagest *of Ptolemy and the Whole of his* Centiloquy (London, 1917)

Ball, Rouse, *A Short History of Mathematics* (New York, 1960)

Baloch, N.A., *Beruni and his Experiment at Nandana* (Islamabad, 1983)

Baron, Roger, 'Notes sur les variations au XIIe siècle de la triade geométrique, altimetria, planimetria, cosimetria', *Isis* 48 (1957), pp. 30–2

Beaujouan Guy, 'Medieval Science in the Christian West', in René Taton (ed.), *Ancient and Medieval Science, A General History of the Sciences*, 4 vols (London, 1963), vol. 1, pp. 468–531

Benson, R.L., and Constable, G. (eds), *Renaissance and Renewal in the Twelfth Century* (Oxford, 1982)

Bischoff, Bernard, 'Die Überlieferung der technischen Literatur', *Mittelalterische Studien* 3 (1981), pp. 277–97

Bjorkman, E., 'Nordische Personennamen in England', in L. Morsbach (ed.), *Studien zur Englischen Philologie* 37 (1910)

Bliemetzrieder, Franz J.P., *Adelhard von Bath: Blätter aus dem Leben eines englischen Naturphilosophen des 12.Jahrhunderts und Bahnbrechers einer Wiedererweckung der griechischen Antike* (Munich, 1935)

Boase, T.S.R., *Castles and Churches of the Crusading Kingdom* (Oxford, 1967)

Boilot, D.J., 'al-Bīrunī', *EoI*, new edn, 4 vols (Leiden, 1960), vol. 1, pp. 1236–8

Boncompagni, B., *see* Section III above

Bowen, James, *A History of Western Education*, 2 vols, vol. 2, *Civilization of Europe Sixth to Sixteenth Century* (London, 1975)

Bradbury, Jim, 'Greek Fire in the West', *History Today* 29 (May, 1979)

Branner, Robert, 'Villard de Honnecourt, Archimedes and Chartres' (USA) *Journal of the Society of Architectural Historians* 19 No. 3 (1960) pp. 91–96

Briggs, Martin S., 'Building Construction', in C. Singer *et al.* (eds), *A History of Technology*, vol. 2, *The Mediterranean Civilization and the Middle Ages* (London, 1956), pp. 397–446

Brooke, Christopher, *The Twelfth Century Renaissance* (London, 1969)

Bubnov, N., *see* Section IV above, Gerberti

Buchtal, Hugo, 'Miniature Painting in the Latin Kingdom of Jerusalem', *Studies in the History of Medieval Secular Illustration* (London, 1971)

Bulmer-Thomas, Ivor, 'Euclid and Medieval Architecture', *Archaeological Journal* 136 (1979), pp. 136–50

Burnett, Charles, 'Adelard, Music and the Quadrivium', *WIST* XIV, pp. 69–86

Burnett, Charles, 'Adelard, Ergaphalau and the Science of the Stars', *WIST* XIV, pp. 133–45, Appendix, *An edition of* Ut testatur Ergaphalau

Burnett, Charles, 'Arabic into Latin in Twelfth-century Spain: The Works of Hermann of Carinthia', *Mittellateinisches Jahrbuch* 13 (1978), pp. 104–106

Burnett, Charles, 'Adelard of Bath and the Arabs', in *Rencontres de Cultures dans la Philosophie Médiévale, Traductions et traducteurs de l'antiquité tardive au XIIe* siècle (Louvain, Cassino, 1990), pp. 89–107

Burnett, Charles, 'Ocreatus', in M. Folkerts and J.P. Hogendijk (eds), *Vestigia Mathematica*, Studies in medieval and early modern mathematics in honour of H.L.L. Busard (Amsterdam, 1993), pp. 69–78

Burnett, Charles, *see also* Sections II, III and IV above

Burnett, Charles, and Cochrane, Louise, 'Adelard and the *Mappae clavicula*', *WIST* XIV, pp. 29–32

Busard, H.L.L., *see* Section III above

Cahen, Claude, *La Syrie du nord à l'époque des Croisades et la principauté franque d'Antioche* (Paris, 1940)

Carmody, Francis J., *Arabic Astronomical and Astrological Sciences in Latin Translation* (Berkeley, 1956)

Chamerlat, Christian Antoine de, *Falconry and Art* (London, 1987)

Chaucer, Geoffrey, *Treatise on the Astrolabe*, ed. W.W. Skeat, Early English Text Society 16 (1872)

Chenu. M.D., *Nature, Man and Society in the Twelfth Century* (Chicago, 1968) (trans. of J. Vrin, *La théologie au douzième siècle*, Paris, 1957)

Childs, W.J., *Across Asia Minor on Foot* (Edinburgh and London, 1917)

Clagett, Marshall, 'Adelard of Bath', *DSB*, vol. 1, pp. 61–4. *See also* Section III above

Clanchy, M.T., *From Memory to Written Record 1066–1307* (London, 1979)

Clanchy, M.T., *England and its Rulers 1066–1272* (Glasgow, 1983)

Cochrane, Louise, 'Adelard of Bath and the Astrolabe', *PSANHS* 124 (1980), pp. 147–50

Colchester, L.S., 'Dimensions and Design', *FWCR* (1981), pp. 15–19

Colchester, L.S. (ed.), *Wells Cathedral. A History* (Open Books, 1982)

Colchester, L.S. and Harvey, John, H., 'Wells Cathedral', *The Archaeological Journal* 131 (1974), pp. 200–14

Conant, K.J., *Carolingian and Romanesque Architecture from 800–1200* (Harmondsworth, 1959)

Cornford, F.M., *see* Section IV above, Plato

Cox, Harding, and Lascelles, Gerald, *Coursing and Falconry* (London, 1892)

Crombie, Alistair, *From Augustine to Galileo*, 2 vols (London, 1951)

Crombie, Alistair, 'Science', ch. 18, in A.L. Poole (ed.), *Medieval England*, 2 vols (Oxford, 1958), vol. 2, pp. 571–604

Crombie, Alistair, 'Quantification in Medieval Physics', *Isis* 52 (1961), pp. 143–160

Crombie, Alistair, 'Some Attitudes to Scientific Progress: Ancient, Medieval and Early Modern', ch. 2, *Science, Optics and Music in Medieval and Early Modern Thought* (London and Ronceverte, West Virginia, 1990), pp. 23–40

Cronne, H.A., and Davis, R.H.C. (eds), *Regesta regum anglo-normannorum 1066–1154*, 4 vols, vol. 3 (Oxford, 1968)

Culpeper, Nicholas, *The Complete Herbal to which are added the English Physician and Key to Physic*, privately printed for Imperial Chemical Industries (London, 1953)

Cunliffe, Barry, *Roman Bath Discovered* (London, 1971)

Curry, Patrick (ed.), *Astrology, Science and Society* (Woodbridge, Suffolk, and Wolfeboro, New Hampshire, 1987)

Curtze, M., *see* Section III above

Cutler, Allan, 'Peter the Venerable and Islam', review of James Kritzeck, *Journal of American Oriental Society* 86 (1966), pp. 184–198

Daniel, Norman, *The Arabs and Medieval Europe* (London, 1975)

Dannenfeld, Karl, H., 'Hermes Trismegistus', *DSB*, vol. 6, pp. 305–306

Davis, H.W.C., *England under the Normans and Angevins* (London, 1949)

Destombes, Marcel, 'Un astrolabe carolingien et l'origine de nos chiffres arabes', *Archives Internationales d'Histoire des Sciences* 58, 59 (1962), pp.3–45

Dickinson, F.A., 'The Sale of Combe', *PSANHS* N.S. 2 (1876)

Dickinson, F.A., 'The Banwell Charters', *PSANHS*, N.S. 3 (1877)

Douglas, David C., *The Norman Fate 1100–1154* (London, 1976)

Drew, Alison, 'The *De eodem et diverso*', *WIST* XIV, pp. 17–24

Dreyer, J.L.E., *History of the Planetary System from Thales to Kepler* (Cambridge, 1906)

Dronke, Peter, *Fabula, Explorations in the Use of Myth in Medieval Platonism* (Leiden, 1974)

Dronke, Peter, *Bernardus Sylvestris Cosmographia* (Leiden, 1978)

Dugdale, W., *see* Section I above

Duhem, P., *Le Système du monde* (Paris, 1915), vol. 3, pp. 112–25

Dunlop, D.M., *Arab Civilization to A.D. 1500* (London, 1971)

Dunning, Robert, *Somerset and Avon* (Edinburgh, 1980)

Epstein, Hans, reviewing Gunnar Tilander, *Grisofus Medicus, Alexander Medicus* (Lund, 1964), in *Speculum* 40 (1965), p. 759

Evans, Gillian, 'From Abacus to Algorism', *British Journal of the History of Science* 16 (1977), pp. 114–31

Evans, Gillian, 'Theory and Practice in Treatises on the Abacus', *Journal of Medieval History* 3 (1977), pp. 21–38

Evans, Gillian, 'A Commentary on Boethius –

Arithmetica of the Twelfth or Thirteenth Century', *Annals of Science* 35 (1978), pp. 131–41

Evans, Gillian, 'Schools and Scholars, the Study of the Abacus', *EHR* 94 (1979), pp. 71–89

Evans, Gillian, *Old Arts and New Theology* (Oxford, 1980)

Evans, Gillian, 'A Note on the Regule abaci', *WIST* XIV, pp. 33–5

Evans, Joan, *Magical Jewels of the Middle Ages and the Renaissance particularly in England* (Oxford, 1922)

Farrer, W., *An Outline Itinerary of King Henry First* (Oxford, 1920; repub. from *EHR* 34 [1919])

Feilitzen, Olaf von, *The Pre-Conquest Personal Names of Domesday Book* (Uppsala, 1937)

Field, J.V., 'Astrology in Kepler's Cosmology', in Patrick Curry (ed.), *Astrology, Science and Society* (Woodbridge, Suffolk, and Wolfeboro, New Hampshire, 1987), pp. 143–70

Finn, Rex W., *Domesday Book, A Guide* (London and Chichester, 1973)

Folkerts, Menso, 'Adelard's Versions of Euclid's *Elements*', *WIST* XIV, pp. 55–68. *See also* Section III above, Adelard's Euclid and Section IV, 'Boethius'

Forssner, T., *Continental Germanic Personal Names in Old and Middle English* (Uppsala, 1916)

Förstemann, E., *Altdeutsches Namenbuch* (Bonn, 1900)

Frankl, Paul, 'The Secret of Medieval Masons', *Art Bulletin of New York* 27 (1945), pp. 45–60

Gandz, Solomon, 'The Origin of the Ghubār Numerals or the Arabian Abacus and the Articuli', *Isis* 16 (1931), pp. 393–424

Gervase of Canterbury, *Historical Works*, ed. William Stubbs (London, 1897), vol. 1, pp. 109–25

Ghyka, Mahla, *Geometrical Composition and Design* (London, 1952)

Gibb, Sharon and De Solla Price, 'An International Checklist of Astrolabes', *Archives Internationales d'Histoire de Science* 30 (1955), pp. 243–63, 363–81

Gibson, Margaret, *Lanfranc of Bec* (Oxford, 1978)

Gibson, Margaret, 'Adelard of Bath', *WIST* XIV, pp. 7–16

Gilson, Etienne Henri, *History of Christian Philosophy in the Middle Ages* (New York, 1955)

Gimpel, Jean, trans. Teresa Waugh, *The Cathedral Builders* (Salisbury, 1983, new edn 1988), pp. 75–97, trans. Carl F. Barnes, Jr (New York, London, 1961), pp. 107–45

Gimpel, Jean, *The Medieval Machine, the Industrial Revolution of the Middle Ages* (London, 1977)

Gleason, S.E., *An Ecclesiastical Barony of the Middle Ages: The Bishopric of Bayeux 1066–1204* Harvard Historical Monographs (Cambridge, Mass., 1936)

Gollancz, H., *see* Section III above, *Quaestiones naturales*

Gombrich, E.H., *The Story of Art* (London, 1957)

Gooder, Eileen A., *Latin for Local History* (London, 1961)

Grant, Edward, *A Source Book of Medieval Science* (Cambridge, Mass., 1974)

Green, Judith, A., *The Government of England under Henry I* (Cambridge, 1986)

Grieve, M., *A Modern Herbal* (London, 1931)

Gunther, R.J. (ed.), *Early Science in Oxford*, 10 vols (Oxford, 1929), vol. 5. (Treatises on the astrolabe by Chaucer and Messahala are reproduced in this volume.)

Gunther, R.J., *Astrolabes of the World*, 2 vols (Oxford, 1932)

Haddon, John, *Bath* (London, 1973)

Hall, Hubert, *The Antiquities and Curiosities of the Exchequer* (London, 1891)

Hartner, Willi, 'Asturlab', *EoI*, vol. 1 (1960), pp. 722–8

Harvey, E. Ruth, 'Qustā ibn Luqā al-Balabakki', *DSB*, vol. 11 (1975) pp. 244–6

Harvey, John, H., 'The Mason's Skill', in Joan Evans (ed.), *The Flowering of the Middle Ages* (London, 1966), p. 177

Harvey, John, H. 'The Origins of Gothic Architecture', *Antiquaries Journal* 48 (1968), pp. 91–94

Harvey, John, H., *The Mediaeval Architect* (London, 1972)

Harvey, John, H., *The Cathedrals of England and Wales* (London, 1974)

Harvey, John, H., 'Wells and Early Gothic', *FWCR* (1978), pp. 19–24

Harvey, John, H., *Mediaeval Gardens* (London, 1981, rev. edn, 1990)

Harvey, John, H., 'Geometry and Gothic Design', *Transactions of the Ancient Monuments Society* 30 (1986), pp. 47–56

Haskins, C.H., 'Adelard of Bath', *EHR* 26 (1911), pp. 491–8

Haskins, C.H., *The Normans in European History* (London, Boston, New York, 1913)

Haskins, C.H., 'Adelard of Bath and Henry Plantagenet', *EHR* 28 (1913), pp. 515, 516

Haskins, C.H., 'The Reception of Arabic Science in England', *EHR* 30 (1915), pp. 56, 57

Haskins, C.H., *Norman Institutions, Harvard Historical Studies* vol. 24 (Cambridge, Mass., 1918, repub. New York, 1960)

Haskins, C.H., 'King Harold's Books', *EHR* 37 (1922), pp. 398–400

Haskins, C.H., *Studies in the History of Mediaeval Science* (Cambridge, Mass. and London, 1927, repub. New York, 1960)

Haskins, C.H., *Studies in Mediaeval Culture* (Oxford, 1929)

Haskins, C.H., and Lockwood, D.P., 'Sicilian Translators of the Twelfth Century and the First Latin Version of Ptolemy's Almagest', *Harvard Studies in Classical Philology* 21 (Cambridge, Mass., 1910), pp. 75–102

Hauréau, J.A., *Histoire de la Philosophie scolastique* (Paris, 1850)

Hay, Denys, *The Medieval Centuries* (London, 1953)

Hay, Denys, 'The Concept of Christendom', in David Talbot Rice (ed.), *The Dark Ages* (London, 1965), pp. 327–43

Heath, Sir Thomas, *see* Section IV above, Euclid

Hermes, Eberhard, trans. P.R. Quarrie, *see* Section IV above, Alfonsi

Hoar, Frank, *An Introduction to English Architecture* (London, 1963)

Hooper, A., *Makers of Mathematics* (London, 1956)

Hunt, Tony, *Plant Names of Medieval Europe* (Cambridge, 1989)

Hunt, W., 'John of Villula', *DNB*, p. 1084

Hunt, W., *The Somerset Diocese, Bath and Wells* (London, 1885). *See also* Section I above

Ibn al Qualanisi, *Continuation of the Chronicle of Damascus, The Damascus Chronicle of the Crusades*, sel. and trans. into English H.A.R. Gibb (London, 1932)

Irani, Rita, 'Arabic Numeral Forms', *Centaurus* 4 (1956), pp. 1–122

James, John, *Chartres, the Masons Who Built a Legend* (London, 1982)

Jervoise, E., *The Ancient Bridges of Mid and Eastern England* (London, 1932), pp. 142, 143

Johnson, Charles, *see* Section IV above, Fitznigel

Jolivet, Jean, 'Adélard de Bath at l'amour des choses', *Metaphysique histoire de la philosophie*, Receuil d'études offert à Fernand Brunner (Neuchâtel, 1981), pp. 77–84

Jones, J.R. and Keller, J.E., *see* Section IV above, Alfonsi

Jones, L.W., 'Cassiodorus', in Austin P.E. Evans (ed.), *Records of Civilization, Sources and Studies* (New York, 1946), pp. 161, 162 on Porphyry

Jones, Peter Murray, *Medieval Medical Miniatures* (London, 1984)

Juschkewitsch, A.P., *see* Section III above, *De numero Indorum*

Keble-Martin, W., *The Concise British Flora in Colour* (London, 1965)

Kealey, E.J., *Roger of Salisbury: Viceroy of England* (Berkeley, 1972)

Kealey, E.J., *Medieval Medicus* (London and Baltimore, 1981)

Kealey, E.J., *Harnessing the Wind* (Woodbridge, Suffolk and Wolfeboro, New Hampshire, 1987)

Kennedy, E.S., 'A Survey of Islamic Astronomical Tables', *TAPS*, N.S. 46 (Philadelphia, 1956)

Kennedy, E.S., 'Late Medieval Planetary Theory', *Isis* 57 (1966), pp. 365–78

Kennedy, E.S., 'Exact Sciences in Iran under the Seljuks and Mongols', in J.A. Boyle (ed.), *Cambridge History of Iran* (Cambridge, 1968) vol. 5, pp. 659–79

King, David A., 'Medieval Astronomical Instruments: A Catalogue in Preparation', *Bulletin of the Scientific Instruments Society* 31 (1991) pp. 3–7

King, David A., 'The earliest known European astrolabe in the light of other early astrolabes', to appear in Wesley Stevens (ed.), *The Oldest Latin Astrolabe* (Proceedings of Session 20 of the XIXth International Congress of History of Science, Saragossa, 22–29 August, 1993 in preparation), Institut für Geschichte der Naturwissenschaften Johann Wolfgang Goethe-Universität Frankfurt am Main (Preprints, 2nd series, 2 January 1994)

Knowles, David, *The Evolution of Medieval Thought* (London, 1962)

Knowles, David, *The Monastic Order in England* (Cambridge, 1963)

Kritzeck, James, *Peter the Venerable and Islam*, Princeton Oriental Studies 23 (Princeton, 1964)

Kunitzsch, Paul, 'Addendum' to E. Poulle, 'Le Traité de l'astrolabe d'Adélard de Bath', *WIST* XIV, p. 131

Labouderie, D., *see* Section IV above, Alfonsi

Latham, R.E., *Revised Medieval Latin Word-List* (London, 1965)

Latham, R.E., and Howlett, D.R., *Dictionary of Medieval Latin from British Sources*, fascicule III, D–E (London, 1986)

Lattin, Harriet, 'The Origin of our Present System of Notation according to theories of Nicholas Bubnov', *Isis* 19 (1933), pp. 181–94. *See also* Section IV above, Gerberti

Lawn, Brian, *The Salernitan Questions* (Oxford, 1963)

Lawn, Brian, *The Prose Salernitan Questions* (London, 1979)

LeClerq, Jean, *The Love of Learning and the Desire for God* (London, 1974)

Leff, Gordon, *Medieval Thought from Augustine to Ockham* (Harmondsworth, 1958)

Legge, M.D., *Anglo-Norman Literature and its Background* (Oxford, 1963)

Legge, M.D., *Anglo-Norman in the Cloisters* (Edinburgh, 1953)

Lemay, Richard, 'The Hispanic Origin of our present Numeral Forms', *Viator* 8 (1977), pp. 435–462

Lemay, Richard, 'Arabic Numerals', *DMA* vol. 1 (1982), pp. 383–98

Lemay, Richard, 'Roman Numerals', *DMA* vol. 10 (1988), pp. 470–74

Lemay, Richard, 'The True Place of Astrology in Medieval Science and Philosophy: Towards a Definition', in Patrick Curry (ed.), *Astrology, Science and Society* (Woodbridge, Suffolk, and Wolfeboro, New Hampshire, 1987), pp. 57–73

Lemay, Richard, 'De la scolastique à l'histoire par le truchement de la philologie: itinéraire d'un médiéviste entre Europe et Islam', estratto dagli atti del Convegno Internazionale promosso dall'Accademia Nazionale dei Lincei-Fondazione Leone Caetani e dall'Universita di Roma 'La Sapienza' Facolta di Lettere – Dipartimento di Studi Orientali, *La difffuzione delle scienze islamiche nel medio evo europeo* (Roma, 2–4 ottobre 1984). Roma (Accademia Nazionale dei Lincei, 1987), pp. 399–533

Lemay, Richard, 'L'authenticité de la Préface de Robert de Chester à sa traduction de *Morienus*', *Chrysopœia*, Tome IV (Paris, 1990–91), pp. 3–32

Lemay, Richard, 'Translators of the Twelfth Century: Literary Issues Raised and Impact Created', in Jeremiah Hackett (ed.), *Medieval Philosophers*, Dictionary of Literary Biography, vol. 115, (Detroit, London, 1992) pp. 367–380

Lemay, Richard, 'De l'antiarabisme – ou rejet du style scolastique – comme inspiration première de l'humanisme italien du Trecento', in Biagio Pelacani Parmense (ed.), *Filosofia, scienza e astrologia nel Trecento europeo* (Parma, 1992)

Lemay, Richard, (in preparation under the aegis of *Instituto Universitario Orientale* of Naples) Abū Ma'shar's *Great Introduction to Astronomy*, edn in 9 vols including the Arabic text and the translations of Abū Ma'shar by John of Seville and Herman of Carinthia

Lemay, Richard, *see also* Section III, above, Abū Ma'shar and Section IV, 'Ptolemy'

Lesser, George, *Gothic Cathedrals and Sacred Geometry*, 3 vols, vol. 3, *Chartres* (London, 1952)

Lewcock, Ronald, 'Architects, Craftsmen and Builders: Materials and Techniques', in George Michell (ed.), *Architecture of the Islamic World* (London, 1978)

Lindberg, David (ed.), *Science in the Middle Ages* (Chicago, 1978)

Lindsay, Jack, *The Normans and Their World* (London, 1973)

Livesey, S.J., and Rouse, R.H., 'Nimrod the Astronomer', *Traditio* 37 (1981), pp. 203–66

Lopez, Robert S., 'Trade of Medieval Europe, the

South', *CEH*, vol. 2, *Trade and Industry in the Middle Ages* (Cambridge, 1952), pp. 257–338, 2nd edn 1987, pp. 306–385.

Lorch, Richard, 'Some Remarks on the Arabic-Latin Euclid', *WIST* XIV, pp. 45–54

Lund, F.M., *Ad Quadratum*, 2 vols (London, 1921)

McLean, Antonia, *Humanism and the Rise of Science in Tudor England* (London, 1972)

McVaugh, Michael, 'Medicine, History of', *DMA*, vol. 8 (1987), pp. 221–54

Maddison, R.E.W., and Maddison, F.R., (trans. of Henri Michel), *Scientific Instruments in Art and History*, 2 vols (London, 1967)

Mahoney, Michael, 'Mathematics', ch. 5, in David Lindberg (ed.), *Science in the Middle Ages* (Chicago, 1978)

Mahoney, Michael, 'Mathematics', *DMA*, vol. 8 (1987), pp. 205–22

Makdisi, J., 'The Scholastic Method in Medieval Education – An Inquiry into its Origins in Law and Theology', *Speculum* 49 (1974), pp. 640–61

Mâle, Emile, *L'art religieux en France du XIIe* siècle (Paris, 1947)

Mâle, Emile, *L'art religieux en France du XIIIe* siècle (Paris, 1948)

Mâle, Emile, *The Gothic Image, Religious Art in France in the Thirteenth Century*, trans. from 3rd edn Dora Nussey (London, 1913, Glasgow, 1961)

Mayer, Leo, *Islamic Astrolabists and Their Work* (Geneva, 1956)

Mercier, Raymond, 'Astronomical Tables in the Twelfth Century', *WIST* XIV, pp. 87–118

Metlitzki, Dorothea, *The Matter of Araby in Medieval England* (New Haven, Conn., 1977)

Michel, Henri, *Traité de l'astrolabe* (Paris, 1947). *See also* Maddison

Michel, P.H., 'Greek Science', in René Taton (ed.), *Ancient and Medieval Science, A General History of the Sciences* 4 vols (London, 1963), vol. 1, pp. 180–239

Migne, J.P. (ed.), *Patrologia Latina*, 221 vols (Paris 1844–64)

Millás, J.M., 'Abū Ma'shar', *EoI*, vol. 1 (1960), p. 139a

Millás-Vallicrosa, J.M., 'La aportacīon astronōmica de Pedro Alfonso', *Estudios sobre historia de la ciencia española* (Barcelona, 1949)

Millás-Vallicrosa, J.M., *Nuevos estudios sobre historia de la ciencia española* (Barcelona, 1960), pp. 105–8

Millás-Vallicrosa, J.M., *see also* Section IV above, Alfonsi

Milne, J., 'Catalogue of Destructive Earthquakes', *British Association for the Advancement of Science*, Report for 1911 (Portsmouth, 1912), Appendix I, pp. 653–5

Mitchell, E.B., *The Art and Practice of Hawking* (London, 1900)

Molland, A.G., 'Medieval Ideas of Scientific Progress', *Journal of the History of Ideas* 39 (1978), pp. 561–77

Moon, Parry, *The Abacus* (New York, London, Paris, 1971)

Moore, Patrick, *The Observer's Book of Astronomy* (London, 1978)

Müller, Martin, *see* Section III, *Quaestiones*

Murdoch, John, 'The Medieval Euclid: Salient Aspects of the Translations of the *Elements* by Adelard of Bath and Campana of Novara', XIIe Congrès international d'histoire des sciences, Colloques (Revue de Synthèse) 3e série, nos. 49–52 (Paris, 1968), pp. 67–94

Murdoch, John, 'Euclid, Transmission of the *Elements*', *DSB*, vol. 4 (1971), pp. 437–59

Nagl, Alfred, 'Über eine Algorismus-schrift des XII. Jahrhunderts und über die Verbreitung der indisch-arabischen Rechenkunst und Zahlreichen im christlichen Abendlande', *Zeitschrift für Mathematik und Physik* 34 (1889), supplement, pp. 129–46, 161–70

Nagl, Alfred, *see also* Section IV above, Ralph of Laon

Nallino, C.A., 'Al Battani', *EoI*, vol. 1 (1960), p. 1104

Nasr, Seyyed Hussein, *Islamic Science, An Introduction to Islamic Cosmological Doctrine* (London, 1976)

Neugebauer, O., 'Early History of the Astrolabe', *Isis* 40 (1949), p. 246

Neugebauer, O., 'Transmission of Planetary Theories in Ancient and Medieval Astronomy', *Scripta Mathematica* 22 (1956)

Neugebauer, O., *Exact Science in Antiquity*, 2nd edn (Providence, Rhode Island, 1957)

Neugebauer, O., 'Studies on Byzantine Astronomical Terminology', *TAPS*, N.S. 50 (1960), part 2

Neugebauer, O., *A History of Ancient Mathematical Astronomy*, 3 parts (Berlin, 1975), part 1, book 1A, *Spherical Astronomy*, pp. 21–52

Neugebauer, O., *see also* Section III, above

Nicholson, Robert Lawrence, *Tancred, a Study of his Career and Work in their Relation to the First Crusade and the Establishment of the Latin States in Syria and Palestine*, University of Chicago Libraries (private edn, 1940)

Nogent, Guibert de, *Autobiographie*, ed. E.R. Labande (Paris, 1989)

Norgate, Kate, 'Matilda' (wife of Henry I), *DNB*, p. 1347

North, John D., 'The Astrolabe', *Scientific American* 230 (January 1974), pp. 96–106 (repub. in *Stars, Minds and Fate, Essays in Medieval Cosmology* (London and Ronceverte, West Virginia,1989), ch. 14, pp. 220, 221; 'Celestial Influences', ch. 18, pp. 243–98)

North, John D., *Richard of Wallingford* (London, 1976)

North, John D., 'Astrology and the Fortunes of Churches', *Centaurus* 24 (1980), pp. 181–213

North, John D., *Horoscopes and History*, *WIST* XIII (London, 1986)

North, John D., 'Some Norman Horoscopes', *WIST* XIV, (1987), p. 159

North, John D., 'Medieval Concepts of Celestial Influence: A Survey', in Patrick Curry (ed.), *Astrology, Science and Society* (Woodbridge, Suffolk, and Wolfeboro, New Hampshire, 1987), pp. 5–17

North, John D., *Chaucer's Universe* (London, 1988)

Norwich, John Julius, *Normans in the South* (London, 1967)

Norwich, John Julius, *Kingdom in the Sun* (Harlow, 1970)

Oggins, Robin and Virginia, 'Richard of Ilchester's Inheritance: An Extended Family in Twelfth-century England', paper read on 7 May 1988 at 23rd International Congress on Medieval Studies, Western Michigan University, Kalamazoo, Michigan

Oggins, Robin and Virginia, 'Some Hawkers of Somerset', *PSANHS* 124 (1980), pp. 51–3

Oswald, Allan, *History and Practice of Falconry* (Jersey, 1982)

Pannekoek, A., *A History of Astronomy* (London, 1961)

Parker, Derek and Julia, *A History of Astrology* (London, 1983)

Patch, Howard, *The Tradition of Boethius* (New York, 1935)

Pelteret, David A. E., *Catalogue of English Post-Conquest Vernacular Documents* (Woodbridge, Suffolk, and Wolfeboro, New Hampshire, 1990)

Phillipps, Sir Thomas, and Way, Albert, *see* Section III above

Pingree, David, 'Astronomy and Astrology in India and Iran', *Isis* 54 (1963), p. 229

Pingree, David, 'Abū Ma'shar', *DSB*, vol. 1 (1970), pp. 35–39

Pirenne, Henri, 'Northern Towns and Their Commerce', *CHME*, vol. 6, *The Victory of the Papacy* (Cambridge, 1929), pp. 505–526

Plato, *see* Section IV above

Plenderleith, R.W., 'The Discovery of an Old Astrolabe', *The Scottish Geographical Magazine* 76, 1 (1960), p. 25

Poole, A.L., *The Oxford History of England*, ed. Sir George Clark, 15 vols, vol. 3, *From Domesday Book to Magna Carta*, 2nd edn (Oxford, 1955)

Poole, A.L. (ed.), *Medieval England* (Oxford, 1958)

Poole, Reginald, *History of the Exchequer in the Twelfth Century* (Oxford, 1912)

Poole, Reginald, *Illustrations of the History of Medieval Thought* (London, 1920)

Poole, Reginald, 'The Masters of the Schools at Paris and Chartres in John of Salisbury's Time', *EHR* 35 (July 1920), pp. 321–42

Poole, Reginald, *Studies in Chronology and History*, coll. and ed. A.L. Poole (Oxford, 1934)

Potter, K.R. (ed. and trans.), *see* Section IV above, *Gesta Stephani*

Poulle, Emmanuel, *Les instruments de la théorie des planètes selon Ptolemée, Equatorie et horlogerie planétaire du XIIIe au XVIe siècles* (Paris, 1980)

Poulle, Emmanuel, 'Le traité de l'astrolabe d'Adélard de Bath', *WIST* XIV, pp. 119–32

Prawer, Joshua, *Histoire du Royaume Latin de Jerusalem* (Paris, 1969)

Previté-Orton, C.W., 'The Italian Cities Till c. 1200', *CHME*, vol. 5, *Contest of Empire and Papacy* (Cambridge, 1926), pp. 208–41

Price, Derek de Solla, *The Equatorie of the Planets* (Cambridge, 1955)

Price, Derek de Solla, with Sharon Gibb, 'An International Checklist of Astrolabes', *Archives Internationales d'Histoire de la Science* 30 (1955), pp. 243–63, 363–81

Price, Derek de Solla, 'Clockwork before the Clock', *Horological Journal* (1956)

Ptolemy, Claudius, *see* Section IV above

Pullen, J.M., *History of the Abacus* (London, 1968)

Reaney, P.H., *A Dictionary of British Surnames* (London, 1958)

Rekaya, M., 'al-Ma'mūn', *EoI*, vol. 6 (1990), 331a–339b

Robbins, F.E., *see* Section IV above, Ptolemy

Robertson, J. Armitage, 'Some Early Somerset Archdeacons', *Somerset Historical Essays* (British Academy, London, 1921)

Rodwell, W., *Excavations and Discoveries, FWCR* (1980)

Rodwell, W., *The Archaeology of the English Church* (London, 1981)

Rosenfeld, B.A., and Grigorian, A.T., 'Thābit ibn Qurra', *DSB*, vol. 13 (1976), pp.288–95

Rowley, Trevor, *The Norman Heritage, 1066–1200* (London, Boston, Melbourne, 1983)

Runciman, Sir Steven, *History of the Crusades*, vol. 2, *The Kingdom of Jerusalem and the Frankish East 1100–1181* (Cambridge, 1952)

Sabra, A.I., 'The Scientific Enterprise', ch. 7, in Bernard Lewis (ed.), *The World of Islam* (London, 1976), pp. 181–200

Sabra, A.I., 'al-Khwārizmī', *EoI* 4 (1978), p. 1068b

Salvin, F.H., and Brodrick, W., *Falconry in the British Isles* (London, 1855, repr. 1980), pp. 126, 127

Sandys, John Edwin, *A Study of Classical Scholarship*, 2 vols (Cambridge, 1903)

Sarton, George, *Introduction to the History of Science*, vol. 1 (Baltimore, 1927), p. 533; vol. 2, part 1 (Baltimore, 1931), pp. 10, 18, 122, 167–9, 209, 210

Sayili, A., *The Observatory in Islam*, Publications of the Turkish Historical Society, series 7, 38 (Ankara, 1960)

Searle, W.G., *Onomasticon Anglo-Saxonicum* (Cambridge, 1897)

Segal, J.B., *Edessa, the Blessed City* (Oxford, 1970)

Shelby, Lon R., 'Geometrical Knowledge of Medieval Master Masons', *Speculum* 47 (1972), pp. 395–421

Silverstein, Theodore, 'Adelard, Aristotle and the *De Natura Deorum*', *Classical Philology* 47, 48 (1952–3), pp. 82–5

Simson, Otto von, *The Gothic Cathedral* (London, 1956), appendix by E. Levy

Singer, Charles, *A Short History of Scientific Ideas* (Oxford, 1959)

Singleton, Barrie, 'Proportions in the Design of the Early Gothic Cathedral at Wells', *British Architectural Association Conference Transactions: Medieval Art and Architecture at Wells Cathedral* (1978)

Smail, R.C., *Crusading Warfare* (Cambridge, 1966)

Smith, David E., *History of Mathematics*, 2 vols (New York, 1958)

Smith and Hawthorne, *see* Section III above, *Mappae clavicula*

Sorabji, Richard, *Time, Creation and the Continuum* (London, 1983)

Southern, Sir Richard, *Western Views of Islam in the Middle Ages* (London, 1962)

Southern, Sir Richard, *Saint Anselm and His Biographer* (Cambridge, 1963)

Southern, Sir Richard, *Medieval Humanism and Other Studies* (Oxford, 1970)

Southern, Sir Richard, *Western Society and the Church in the Middle Ages*, vol. 2 , *The Pelican History of the Church*, ed. O. Chadwick (Harmondsworth, 1970)

Southern, Sir Richard, 'The Schools of Paris and the School of Chartres', in R.L. Benson and G. Constable (eds), *Renaissance and Renewal in the Twelfth Century* (Oxford, 1982) pp. 113–37

Stenton, F.M., *William the Conqueror and the Rule of the Normans* (London, 1908)

Stenton, F.M., *The First Century of English Feudalism 1066–1166* (Oxford, 1932)

Stiefel, Tina, 'Science, Reason and Faith in the Twelfth Century: The Cosmologists' Attack on Religion', *Journal of European Studies* 6 (1976), pp. 1–16

Stiefel, Tina, 'The Heresy of Science, A Twelfth-century Conceptual Revolution', *Isis* 68 (1977), pp. 347–62

Stiefel, Tina, 'Twelfth-century matter for metaphor: the material view of Plato's *Timaeus*', *British Journal for the History of Science* 17 (1984), pp. 169–75

Stiefel, Tina, *The Intellectual Revolution in Twelfth-Century Europe* (New York, 1985)

Stock, Brian, *Myth and Science in the Twelfth Century: A Study of Bernard Sylvester* (Princeton, 1972)

Stratford, Neil *et al.*, 'Archbishop Hubert Walter's Tomb and its Furnishings', *Medieval Art and Architecture in Canterbury* (British Archaeological Association, 1982), pp. 87–93

Stratford, Neil, 'Metalwork', *English Romanesque Art 1066–1200: Catalogue of the Exhibition* (Hayward Gallery, London, 1984), pp. 232–95

Suter, H., 'Al Badi al Asturlabi', *EoI*, vol. 1 (1960), p. 858. *See also* Section III above, al-Khwārizmī

Sweet, Henry, 'The Oldest English Texts', *Early English Text Society* 83 (1885)

Talbot Rice, David, *Islamic Art* (London, 1965)

Tardif, E.J., *Coutumiers de Normandie* (Paris, 1881, repr. Geneva, 1977)

Thompson, J.W., 'The Introduction of Arabic Science into Lorraine in the Tenth Century (984)', *Isis* 12 (1927), pp. 184–93

Thomson, Ron, 'Jordanus de Nemore and the Mathematics of Astrolabes', *Pontifical Institute of Mediaeval Studies* 39 (Toronto, 1978)

Thorndike, Lynn, *A History of Magic and Experimental Science*, 2 vols (New York, 1923)

Thorndike, Lynn, 'Chiromancy', *Speculum* 40 (1965), pp. 674–706

Thorndike, Lynn and Kibre, Pearl, *see* Section III above

Tilander, Gunnar, *Grisofus Medicus, Alexander Medicus* (Lund, 1964)

Tomkis, Thomas, *Albumazar, A Comedy* (1615), ed. Hugh G. Dick (Berkeley, 1944)

Toomer, G.J., review of O. Neugebauer, *'Zij* of al-Khwārizmī', *Centaurus* 10 (1964), p. 203

Toomer, G.J., 'Al-Khwārizmī', *DSB*, vol. 7 (1973), pp. 358–65

Toomer, G.J., *see also* Section IV above, Ptolemy

Usher, George, *A Dictionary of Plants Used by Man* (London, 1974)

Venables, Edmund, 'Robert Losinga (*de Lotharingia*)' (Bishop of Hereford), *DNB*, vol. 1, p. 1245

Victor, Stephen K., *Practical Geometry in the High Middle Ages, Memoir, APS*, vol. 134 (1979)

Villard de Honnecourt, *Carnet*, Introduction et commentaires de Alain Erlande-Brandenburg, Regine Pernoud, Jean Gimpel, Roland Bechmann (Paris, Editions Stock, 1986)

Vitruvius, *The Ten Books of Architecture*, trans. Morris Hicky Morgan (Harvard University Press, 1914, repub. Dover Publications, New York, 1960)

Vogel, K., 'Theophilus to the Fourth Crusade', in J.M. Hussey (ed.) with assistance of D.M. Nicol and G. Cowan, *CHME*, vol. 4, part 2, *Byzantine Science* (Cambridge, 1967), pp. 270–83. *See also* Section III above

Waddell, Helen, *The Wandering Scholars*, 6th edn (London, 1932)

Waters, D.W., *The Planispheric Astrolabe*, National Maritime Museum (Greenwich, 1976)

Wedel, Theodore, *The Mediaeval Attitude towards Astrology particularly in England*, Yale Studies in English 60 (New Haven, USA, 1920)

Welborne, M.C., 'Lotharingia as a center of Arabic and scientific influence', *Isis* 16 (Cambridge, Mass., 1931), pp. 188–99

William of Malmesbury, *History of the Kings of England*, trans. Revd John Sharpe (London, 1815)

Wright, J.K., 'Notes on the knowledge of latitudes and longitudes in the middle ages', *Isis* 5, part 1 (Brussels, 1923), pp. 75–98

Wright, J.K., *Geographical Lore at the Time of the Crusades* (New York, 1965)

Wright, T., *Biographia Britannica Litteraria*, vol. 2 (1896), p. 94

Yates, F.A., *Giordano Bruno and the Hermetic Tradition* (London, 1964)

Yeldham, Frances, 'Fraction Tables of Hermann Contractus', *Speculum* 3 (1928), pp. 240–5

Index